VOLUME ONE HUNDRED AND FOURTEEN

ADVANCES IN
COMPUTERS

VOLUME ONE HUNDRED AND FOURTEEN

ADVANCES IN
COMPUTERS

Edited by

ALI R. HURSON
Missouri University of Science and Technology, Rolla,
MO, United States

ACADEMIC PRESS
An imprint of Elsevier

Academic Press is an imprint of Elsevier
50 Hampshire Street, 5th Floor, Cambridge, MA 02139, United States
525 B Street, Suite 1650, San Diego, CA 92101, United States
The Boulevard, Langford Lane, Kidlington, Oxford OX5 1GB, United Kingdom
125 London Wall, London, EC2Y 5AS, United Kingdom

First edition 2019

Notices
Knowledge and best practice in this field are constantly changing. As new research and experience broaden our understanding, changes in research methods, professional practices, or medical treatment may become necessary.

Practitioners and researchers must always rely on their own experience and knowledge in evaluating and using any information, methods, compounds, or experiments described herein. In using such information or methods they should be mindful of their own safety and the safety of others, including parties for whom they have a professional responsibility.

To the fullest extent of the law, neither the Publisher nor the authors, contributors, or editors, assume any liability for any injury and/or damage to persons or property as a matter of products liability, negligence or otherwise, or from any use or operation of any methods, products, instructions, or ideas contained in the material herein.

ISBN: 978-0-12-817157-8
ISSN: 0065-2458

For information on all Academic Press publications
visit our website at https://www.elsevier.com/books-and-journals

Publisher: Fiona Pattison
Acquisition Editor: Fiona Pattison
Editorial Project Manager: Peter Llewellyn
Production Project Manager: James Selvam
Cover Designer: Greg Harris

Typeset by SPi Global, India

Working together
to grow libraries in
developing countries

www.elsevier.com • www.bookaid.org

Contents

PREFACE

Traditionally, *Advances in Computers*, the oldest series to chronicle of the rapid evolution of computing, annually publishes several volumes, each one typically comprised of four to eight chapters, describing new developments in the theory and applications of computing. The 114th volume is an eclectic volume inspired by several issues of interest in research and development in computer science. This volume is a collection of four chapters that were solicited from authorities in the field, each of whom brings to bear a unique perspective on the topic.

Due to its high performance and density flash memory-based solid-state drives (SSDs) are becoming ubiquitous in modern computing systems. And, they are widely used as a secondary memory because of their superior performance in comparison to magnetic hard disk drives. In Chapter 1, "A comprehensive survey of issues in solid state drives," Youngbin Jin and Ben Lee present a comprehensive study of SSDs topics that span across from the physical properties of a flash memory cell to the architecture of SSDs. In addition, it addresses recent research efforts on SSDs.

In Chapter 2, "Revisiting VM performance & optimization challenges for big data," Nayyer et al. articulate the concept of virtualization in cloud computing as a means to maximize resource utilization and minimize cost by deploying multiple virtual machines on a single physical platform. However, sharing computing resources by multiple virtual machines can result in severe performance degradation, motivating virtual machine migration techniques. Introduction of big data brings out new challenges to the existing virtual machines performance enhancement. This chapter describes big data-triggered VM performance challenges focusing on big data applications and storage migration in cloud computing. State-of-the-art VM migration techniques are evaluated against challenges posed by big data to motivate possible solutions and research challenges.

In Chapter 3 entitled, "Toward realizing self-protecting healthcare information systems: Design and security challenges," Qian Chen studies the history of Healthcare Information Systems (HISs) and discusses recent cyber security threats affecting HISs and then introduces the autonomic computing concept. This chapter also analyzes the security vulnerabilities of the HIS network, communication links, and protocols. Based on these studies the concept of self-protecting HISs (SPHISs) that can defend themselves against cyber intrusions with little or no human intervention is articulated. This chapter articulates that a SPHIS should contain monitoring systems, early estimation modules, intrusion detection, network forensics analysis devices, and intrusion response systems. Finally, existing self-protecting approaches for HIS, enterprise systems, and industrial control systems are demonstrated in detail.

Finally, in Chapter 4, "SSIM and ML based QoE enhancement approach in SDN context," Asma Ben Letaifa is looking at a new performance metric, i.e., quality of experience within the framework of video streaming. This chapter first articulates the so-called QoS, QoE, mobile cloud computing environment, and software-defined network. Mathematical techniques in modeling, predicting, and evaluating QoE are presented. Finally, a machine learning adaptive coding to provide a better QoE for video streaming services is presented and evaluated.

I hope that readers find these articles of interest, and useful for teaching, research, and other professional activities. I welcome feedback on this volume, as well as suggestions for topics of future volumes.

ALI R. HURSON

Missouri University of Science and Technology,

Rolla, MO, United States

CHAPTER ONE

A comprehensive survey of issues in solid state drives

Youngbin Jin, Ben Lee
School of Electrical Engineering and Computer Science, Oregon State University, Corvallis, OR, United States

Contents

Advances in Computers, Volume 114
ISSN 0065-2458
https://doi.org/10.1016/bs.adcom.2019.02.001

Abstract

Flash memory-based solid state drives (SSDs) have become ubiquitous in modern computing systems, such as high-performance servers, workstation, desktops, and laptops, due to their performance and density. The architecture of SSDs has evolved to exploit the advantages of flash memories and at the same time conceal their shortcomings. SSD concurrency techniques such as channel striping, flash-chip pipelining, die interleaving, and plane sharing utilize the available parallelism of flash memories, while the flash translation layer (FTL) operations minimize the latency overhead of flash memories. This chapter provides a comprehensive survey of SSD topics that span across from the physical properties of a flash memory cell to the architecture of SSDs. The FTL-related topics are discussed in the context of inter-related system-level operations, which include address mapping, garbage collection, wear leveling, bad block management, SSD concurrency techniques, and page allocation strategies. This chapter also surveys recent research efforts on SSDs.

1. Introduction

In the recent years, flash memories have become the main storage technology for computers and mobile devices. NAND flash memory-based solid–state drives (SSDs) are widely used as a secondary memory in modern computing systems due to their superior performance compared to magnetic hard disk drives (HDDs). The performance of HDDs has stagnated due to the limitations in the rotational speed of magnetic platters and the seek time of actuator arms. In contrast, SSDs do not have complex mechanical parts, resulting in lower latency as well as lower failure rate than HDDs [1]. Furthermore, SSDs offer superior bandwidth, higher random I/O performance, lower power consumption, higher shock resistance, and improved system reliability compared to HDDs [2, 3].

The first version of SSDs were RAM-based invented by StorageTeK in 1978. Flash memory-based SSDs were introduced in 1989 by Western Digital. Early SSDs were implemented with the NOR flash memory structure due to its high performance. SSDs with the higher density NAND flash memory structure were developed by M-System in 1995; however, their use was limited to certain applications (e.g., in industrial or military settings) due to their high cost. By 2004, the cost of NAND flash memories dropped drastically, and SSDs entered a new phase as a replacement for magnetic storage devices. Flash memories have since been used as storage for portable devices, such as smartphones, pad devices, laptops, etc. They are also used in desktops and high-performance servers due to their high capacity and bandwidth [2, 4].

Despite the popularity and advantages of SSDs, flash memories have critical shortcomings. First, a flash memory cell has to be erased first before it can be programmed, referred to as *erase-before-write*, due to its physical characteristics. Furthermore, the erase operation has a long latency that needs to be hidden. Second, the oxide layer in a flash memory cell will become damaged over many program and erase (P/E) cycles. This causes the flash memory cell to become unreliable limiting its lifetime, which is referred to as the *endurance problem* [3].

SSD architectures have been developed to exploit the advantages of flash memories and to hide their disadvantages. In order to achieve these goals, SSDs employ *flash translation layer* (FTL) to perform various operations to maximize their performance, lifetime, and reliability. FTL consists of the following functionalities:

- Maps logical addresses from the host system to physical addresses of flash memories.
- Due to the erase-before-write requirement of flash memory, write requests to valid pages (i.e., updates) must be performed on new blocks and the original pages need to be marked as invalid. This process called *out-of-place updating* increases the number of invalid pages. Therefore, these invalid pages need to be reclaimed for future allocation by performing a background operation referred to as *garbage collection*.
- Since flash memories have a limited number of P/E cycles, their reliability is not guaranteed when the number of P/E cycles exceeds the maximum threshold. Therefore, these P/E cycles need to be managed by performing *wear leveling* to evenly wear out flash blocks [1–3, 5].
- Performs *bad block management* to avoid using blocks that have exceeded the P/E cycle limit and thereby maintain high reliability [6, 7].
- Exploits *concurrency* using techniques such as *channel striping*, *flash-chip pipelining*, *die interleaving*, and *plane sharing* in order to maximize performance.
- Checks the reliability of blocks using *error correction code* (ECC).

Based on the aforementioned discussion, this chapter provides a comprehensive survey of issues in SSDs. Unlike existing surveys that are limited to specific areas of SSD, such as FTL [8, 9], endurance and reliability [10], or architecture [11], this chapter provides a broad coverage of issues from the *Flash device-level* to the *SSD system-level*. Moreover, this chapter also surveys recent research efforts on SSDs.

The chapter is organized as follows: Section 2 discusses the fundamentals of a flash device, such as material composition, the flash memory characteristics that allow "0" and "1" to be distinguished, and inherent endurance limitation. Based on this background, the rest of the section covers the basic operations of a NAND flash memory. Section 3 provides a brief description of the SSD architecture and discusses the structure of a flash memory from one transistor cell to an entire SSD device. The importance of FTL is introduced in Section 4. Afterward, Sections 5–8 cover address mapping, garbage collection, wear leveling, and bad block management, respectively. Section 9 discusses the various SSD parallelism techniques. Section 10 introduces page allocation strategies on SSDs. Recent research trends in SSDs are presented in Section 11. Finally, Section 12 concludes the chapter.

2. Fundamentals of a flash device

SSDs are designed to exploit the advantages of flash devices and at the same time to conceal their disadvantages. For this reason, this section presents the fundamentals of flash memory, including the device structure and basic operations, to better understand flash memory-based SSDs.

2.1 Flash device

A flash memory cell is made up of a metal oxide semiconductor field effect transistor (MOSFET) with a floating gate as shown in Fig. 1. A floating gate is inserted between the metal control gate and the substrate layer. This floating gate allows the flash cell to maintain one of the two possible states, which are distinguished by whether or not there are electrons trapped in the floating gate.

Fig. 2 illustrates the states "1" and "0". When there are no electrons in the floating gate, the state of the cell is considered to be "1". When there are electrons in the floating gate, the state of the cell is considered to be "0". The trapped electrons in the floating gate shift the threshold voltage (V_t) level of the MOSFET. Thus, a floating gate with trapped electrons has a higher V_t

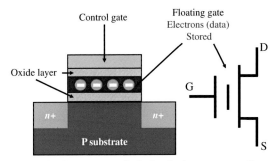

Fig. 1 Cross section and schematic symbol of a flash memory cell.

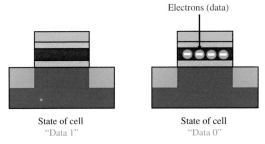

Fig. 2 The states of a flash memory cell.

than a floating gate with no electrons, i.e., $V_t^{data1} < V_t^{data0}$. In order to read the state of a cell, a read reference voltage (V_{READ}) is applied to the control gate, where $V_t^{data1} < V_{READ} < V_t^{data0}$ [12]. Then, a sensing circuit distinguishes whether or not there is current flowing through the cell to read data. If V_t is higher than V_{READ}, there will be no current through the cell. On the other hand, if V_t is lower than V_{READ}, current passes through the cell. Note that the floating gate can retain electrons for a few years without power, and this physical characteristic (i.e., long retention time) allows flash memories to be used as nonvolatile storage [12, 13].

2.2 Basic operations of flash memory

The basic operations of flash memory are program, read, and erase. Fig. 3 illustrates the program and erase operations. The following explains the three operations in more detail:

2.2.1 Program

A *program* operation involves injecting electrons into the floating gate to write a "0". The Fowler–Nordheim (FN) tunneling is used to inject

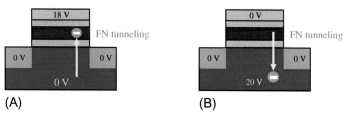

Fig. 3 Flash memory program (A) and erase operations (B).

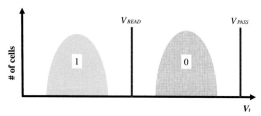

Fig. 4 SLC read voltage levels.

electrons into the floating gate. FN tunneling occurs when a high voltage (e.g., 18 V) is applied across the control gate and the substrate as shown in Fig. 3A [14].

2.2.2 Read

As discussed in Section 2.1, the basic principle of a *read* operation is based on V_t. Fig. 4 shows the V_t distribution of a single-level cell (SLC) flash memory structure that stores one bit of data, where V_t for data "1" (V_t^{data1}) is lower than V_{READ}, while V_t for data "0" (V_t^{data0}) is higher than V_{READ}. V_{PASS} is a voltage level higher than any of the data states to pass the current during a page read operation (see Section 3.2). In order to perform a read operation, V_{READ} is applied to the control gate of a flash cell being read. Then, the stored data can be determined based on whether or not current passes through the transistor. That is, if current is detected, it means the cell contains data "1". Otherwise, the cell contains data "0".

Fig. 5 shows the V_t distribution of a multilevel cell (MLC) flash memory technology, which is divided into four voltage windows to store two bits of data in one flash cell. A read operation involves a two-step process. First, the most significant bit is detected by applying V_{READ2} to distinguish between data sets "1x" and "0x". Afterward, the least significant bit can be detected by applying V_{READ1} to distinguish between data "11" and "10" or V_{READ3} to distinguish between data "01" and "00". Note that a triple-level cell (TLC) flash memory technology can hold three bits of data in one flash cell.

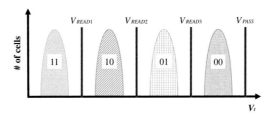

Fig. 5 MLC read voltage levels.

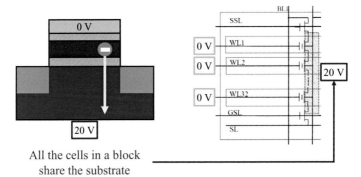

All the cells in a block
share the substrate

Fig. 6 Erasing an entire block.

2.2.3 Erase

An *erase* operation is the process of removing electrons from the floating gate in order to change the state of the cell to "1" [12, 15]. During an erase operation, a large negative voltage is required to repel electrons from the floating gate as shown in Fig. 3B [12, 15]. This is accomplished by grounding the control gate and applying 20 V to the substrate.[a] As a result, electrons are removed from the floating gate due to the FN tunneling effect.

The erase operation is performed on a block-by-block basis, which means that an individual flash cell cannot be changed from "0" to "1", unlike from "1" to "0" as shown in Fig. 6. Thus, new empty (erased) blocks have to be prepared before programming (i.e., erase-before-write). Moreover, the erase operation has a much longer latency than the read and program operations. For example, the read, program, and erase latencies for Micron 8 Gb flash-chip are 25 μs, 220 μs, and 1500 μs, respectively [16]. For this reason, the erase operation is one of the critical performance bottlenecks for NAND flash memories, and many FTL algorithms are designed to conceal the long erase latency (Fig. 6).

---

[a] Note that it is possible to apply −20 V at the control gate. However, this requires an extra charge pump, which adds cost in terms of power consumption and hardware implementation.

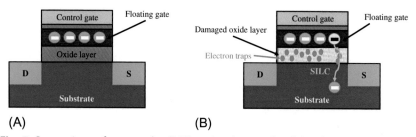

Fig. 7 Comparison of a normal cell (A) and a damaged cell (B) due to repeated P/E operations.

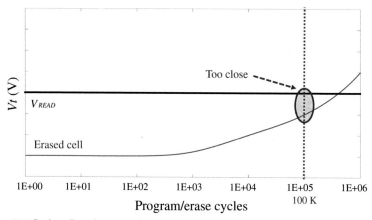

Fig. 8 SLC flash cell endurance characteristics [10].

2.3 Endurance

As mentioned in Section 1, a flash cell has a limited number of P/E cycles. Typical P/E cycles for SLC, MLC, and TLC are 100 k, 10 k, and 1 k, respectively [17]. If these limitations are exceeded, their reliability as flash memories is not guaranteed. This endurance problem is caused by the damaged oxide layer due to repeated P/E operations as shown in Fig. 7 [10, 18–20]. The damaged oxide layer causes (1) electrons to become trapped in the tunnel oxide layer increasing V_t for an erased cell and (2) electrons to escape from the floating gate faster due to stress–induced leakage current (SILC) reducing retention time of a programmed cell. Fig. 8 shows how V_t of an erased cell increases as P/E cycles increase [10]. After V_t crosses the reference voltage V_{READ}, the cell cannot reliably operate as a memory element.

3. SSD architecture and flash memory structure

3.1 SSD architecture

Fig. 9 shows an SSD architecture [2], which consists of Processor, RAM, Host Interface, Buffer Manager, multiple Flash Controllers and Channels, and Flash-Chips. The *Processor* manages most of the SSD operations, such as request/response flow and logical-to-physical address mapping. The *RAM* consisting of SRAM and DRAM supports the functionalities of the Processor. The *Host Interface* provides a physical interface between the host and SSD, such as PCIe, SATA, SAS, etc. A *Flash Controller* handles commands from the Processor and manages the operations of Flash-Chips on its channel [2]. It is also responsible for transferring data between the Flash-Chips and the Buffer Manager. The multiple channels and flash-chips allow for exploitation of I/O parallelism [21] (see Section 9). The main responsibility of the *Buffer Manager* is to hold commands and data in order to handle requests and responses between the Host Interface and the Flash Controller.

Fig. 10 provides a more detailed illustration of a flash-chip in Fig. 9. Each flash-chip consists of dies, planes, blocks, and pages [16]. The *I/O Control* is used to communicate with the Flash Controller. The *Control Logic* issues

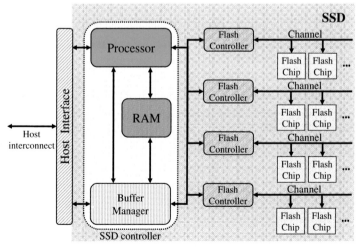

Fig. 9 SSD architecture [2].

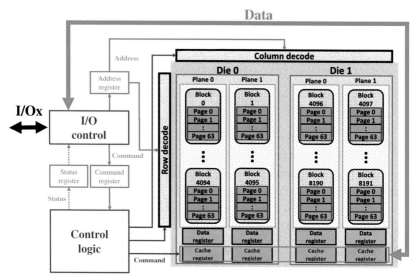

Fig. 10 Micron 8 Gb flash-chip [16].

commands and corresponding addresses to the flash array. *Data/Cache regis-ters* support read/write operations to NAND flash memory.

Sections 3.1.1–3.1.3 explain the SSD read, write and advanced operations.

3.1.1 SSD read operation

An SSD read operation involves three phases: read set (RS), read from NAND (RN), and read data (RD). Fig. 11 illustrates the read operation. In the *RS* phase, a read command and the related address are sent to the Buffer Manager via the Host Interface. In the *RN* phase, the Buffer Manager first sends command/address to the Flash Controller then the Flash Control-ler transfers command to the corresponding flash-chip (i.e., setup). The flash-chip reads data from NAND flash array to the Data/Cache register, and the flash-chip is in busy state during this process. Afterward, the flash-chip transfers data to the Flash Controller, then the Flash Controller moves it to the Buffer Manager (i.e., data transfer). In the *RD* phase, the host system receives data from the Buffer Manager via the Host Interface using direct memory access (DMA) [11, 13].

3.1.2 SSD write operation

An SSD write operation consists of three phases: write set (WS), write data (WD), and write to NAND (WN). The SSD write operation is depicted in Fig. 12. In the *WS* phase, the host system sends a program command and the related address to the Buffer Manager via the Host Interface. Then, the

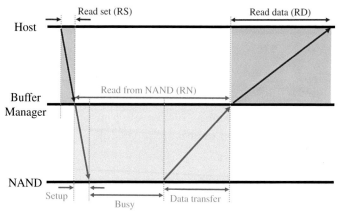

Fig. 11 SSD read operation.

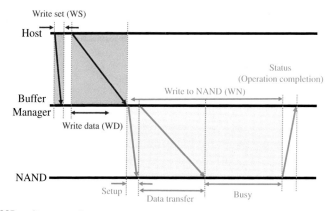

Fig. 12 SSD write operation.

host system transfers data during the *WD* phase. During the *WN* phase, the Buffer Manager issues a program command/address, and then the Flash Controller sends command to the corresponding flash-chip (Setup). Afterward, data are transferred from the Buffer manager to the flash-chip via the Flash Controller (Data transfer). Then, the corresponding flash-chip writes data to its NAND flash array, and the state of this flash-chip is in busy state during the write operation. When the write operation is done, the flash-chip sends the status signal to notify the completion of the write operation [11, 13].

3.1.3 *Advanced operations*

A modern flash-chip provides advanced operations such as *Two-plane Page Read, Two-plane Page Program, Multiple-die Page Read, Interleaved Program Page, Page Read Cache Mode,* and *Internal Data Move* to exploit internal parallelism within a flash-chip [16]. The *Two-plane Page Read/Program* operations allow for the two planes of the same die to be read/programmed at the same time. The *Multiple-die Page Read* and the *Interleaved Program Page* operations support the two dies of a flash-chip to be read and programmed, respectively, at the same time. The *Page Read Cache Mode* operation is used to speedup sequential read operations within a block. Finally, the *Internal Data Move* (copy-back) operation moves one page of data to another page within the same plane using the data register [16].

3.2 Structure of a flash device

There are two major flash memory structures: NAND flash and NOR flash. NOR flash has better read and write speeds than NAND flash. Although NOR flash memories provide higher performance, NAND flash memories are widely used in SSDs due to their small cell size (i.e., a low cost per bit). For this reason, NAND flash memory is mainly discussed in this paper.

A NAND flash-chip consists of pages, blocks, planes, and dies. A *page* consists of a number of flash cells that share the same word line, typically 512 B–16 KB. There is also additional spare area in each page to store a few bytes of management information such as ECC and mapping tables to support FTL operations [22]. The performance of read and program operations improves as the page size increases due to parallelism; however, it is limited by RC delay and lithography. In contrast to the read and program operations, erase operations are performed on a *block* granularity consisting of 32, 64, 128, or 256 pages. A large block size enhances array efficiency since more cells share the select gate and contact area. However, the number of valid data pages that need to be relocated during garbage collection increases, which is referred as *write amplification* (see Section 6) [23]. A *plane* is composed of 1024 or 2048 blocks. Finally, a *die* is made up of 2 or 4 planes, and there are 2 or 4 dies per flash-chip.

Fig. 10 shows an example NAND flash-chip, which is a Micron 8 Gb flash-chip with 2 dies. The capacity of an SSD is determined by the number of flash-chips, i.e., capacity = 8 Gb $\times n$ flash-chips. A flash-chip is composed of 2 dies, where each die contains 4 Gb flash memory. Each die is made up 2 planes, where a plane consist of 2048 blocks. Each block has 64 pages, and the page size is 2 KB [16]. Thus, the relationships for different granularity levels are summarized as follows:

1 page = 2048 + 64 (spare area) bytes
1 block = (2048 + 64) bytes × 64 pages = 132 KB

1 plane = 132 KB × 2048 blocks = 2112 Mb
1 die = 2112 Mb × 2 planes = 4224 Mb
1 flash–chip = 4224 Mb × 2 dies = 8448 Mb
1 SSD device = 8 Gb × n flash–chip

A NAND flash array is shown in Fig. 13 [13], which represents one of the planes in Fig. 10. A NAND string is a serial connection that shares the same bit line (i.e., a vertical line in Fig. 13, where all the transistors are connected through their source and drain), and a NAND page is a serial connection that shares the same word line (i.e., the horizontal line in Fig. 13, where all the transistors connected through their control gates). Since a NAND page shares the same control gate, the read and program operations are performed at a page granularity.

Programming individual cells within a page requires the program and program inhibit states shown in Fig. 14. For example, when data "0" needs to be programmed into the cell at word line (WL) 2 and bit line (BL) 1, i.e., (WL2, BL1), a high voltage is applied to WL2 and BL1 is grounded. Then, the control gate and substrate at the cell (WL2, BL1) has enough voltage difference (18 V) to cause FN tunneling. Meanwhile, the cell (WL2, BL2) is not a target for programming data "0", but it shares the same control gate (i.e., WL2) with the cell (WL2, BL1). Thus, the *program inhibit* state is

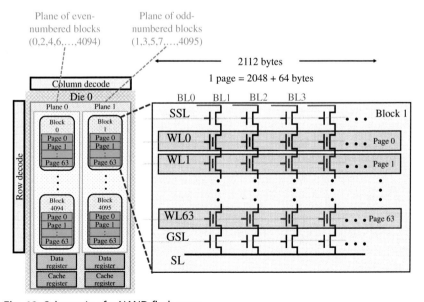

Fig. 13 Schematic of a NAND flash array.

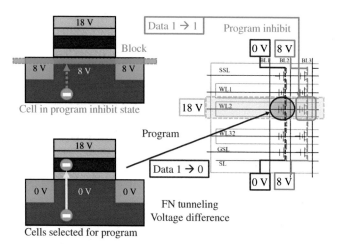

Fig. 14 Program and program inhibit.

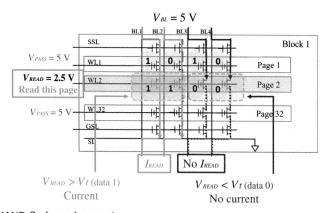

Fig. 15 NAND flash read operation.

required to prevent this cell from being programmed. This is done by applying a voltage (i.e., 8 V) to BL2 as shown in Fig. 14 such that it will not cause FN tunneling. Therefore, the cell (WL2, BL1) will be programmed with data "0" while the data at cell (WL2, BL2) will remain at "1".

As mentioned before, NAND flash read operations are done at a page granularity. In SLC, V_{READ} is applied to all the gates of a page being read, while the gates of the other pages are driven to V_{PASS} (usually 5 V) to act as pass transistors. Then, the stored data can be determined based on whether or not the current passes through the transistor. Fig. 15 illustrates a read operation of a NAND flash array. Suppose the flash memory array has the following data: (WL1, BL1) = 1, (WL1, BL2) = 0, (WL1, BL3) = 1, (WL1, BL4) = 0, (WL2, BL1) = 1, (WL2, BL2) = 1, (WL2, BL3) = 0, and (WL2, BL4) $= 0$. In order to read the target page on WL2 in Fig. 15, V_{READ} (2.5 V) is applied to WL2 and V_{PASS} (5 V) is applied to all other WLs. The cells (WL2, BL1) and (WL2, BL2) allow the current to pass since V_{READ} is higher than V_t of these cells (containing data "1"). On the other hand, the cells (WL2, BL3) and (WL2, BL4) obstruct the current because V_{READ} is lower than V_t of these cells (containing data "0"). Data in other pages (i.e., other WLs) are not affected when sensing the target WL because V_{PASS} is always higher than either data states as indicated in Fig. 15.

4. Flash translation layer

The flash translation layer (FTL) is an intermediate system made up software and hardware that manages SSD operations as shown in Fig. 16 [1]. The FTL performs logical-to-physical address translation, garbage collection, wear-leveling, error correction code (ECC), and bad block management [4, 24]. In addition, FTL exploits data concurrency techniques, such as striping, interleaving, and pipelining, to achieve high throughput. Because of the need to perform garbage collection, wear leveling, and bad block management, the performance of SSDs drops off after all the flash blocks have been written to as shown in Fig. 17. This performance drop is called the *write cliff* [21]. As such, the FTL is crucial to SSD performance, and thus most of the system or architecture level research on SSDs is focused on FTL-related topics.

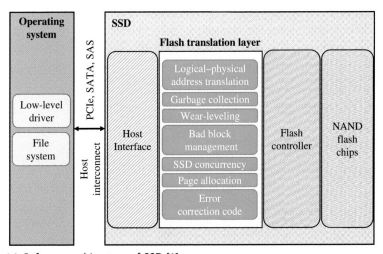

Fig. 16 Software architecture of SSD [1].

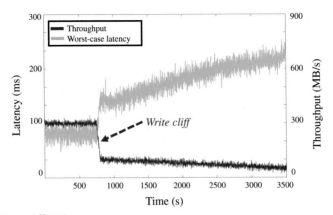

Fig. 17 Write cliff [21].

One of the main functions of FTL is to map logical addresses from the file system to physical addresses of the NAND flash memory. This address mapping will be discussed in Section 5. Afterward, the FTL needs to find new blocks to write data. In particular, update requests need to be written to free blocks due to the out-of-place update requirement and the pages of the original blocks are marked as invalid. As a result, many invalid pages can remain in the blocks. *Garbage collection* reclaims obsolete blocks that have more invalid pages than a given threshold, which will be explained in more detail in Section 6. Since flash memories have limited P/E cycles, *wear-leveling* is needed to evenly wear out flash devices in order to achieve longer lifetime. Wear-leveling techniques will be discussed in detail in Section 7. Despite the use of wear-leveling, flash memory blocks will eventually wear out. Therefore, these worn out blocks, referred to as bad blocks, need to be managed. *Bad block management* avoids the use of bad blocks due to the endurance problem and manufacturing faults. Bad block management techniques will be discussed in Section 8.

5. Address mapping

Address-mapping techniques are classified as page-level mapping, block-level mapping, and hybrid mapping based on the unit of flash memory granularity [8, 24]. Sections 5.1–5.3 discuss each of these address-mapping techniques.

5.1 Page-level mapping

Fig. 18 illustrates the *page-level mapping* scheme, where a logical address is mapped to a physical address at a page granularity. In this example, the file system sends the command *write*(5, *A*), i.e., write data "A" to the logical page number (LPN) 5, and the *page mapping table* maps this to the physical page number (PPN) 10 in Block 2. The page-level mapping technique offers flexibility in choosing any valid pages to perform reads and writes. However, the page mapping table must maintain all the logical to physical mapping information for the entire flash memory, which requires a mapping table size of $\lceil (log_2 P)/8 \rceil \times P$ bytes, where P is the number of pages in the SSD. For instance, a 1 TB SSD with 4 KB pages would require a mapping table with 2^{28} entries each with a 4-byte address. This leads to a mapping table size of 1 GB, which is too large to be stored in an SRAM [8, 9, 24]. Furthermore, the size of the mapping table increases as the size of flash memories increases.

5.2 Block-level mapping

In order to reduce the mapping table size, *block-level mapping* can be employed. Fig. 19 shows the block-level mapping scheme, where an LPN is converted to a physical block number (PBN) and an offset. In the example shown in Fig. 19, LPN 5 is divided by the number of pages per

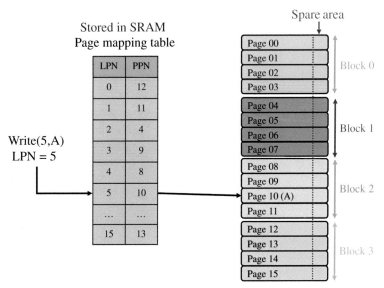

Fig. 18 Page-level mapping.

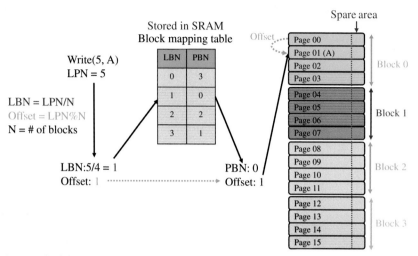

Fig. 19 Block-level mapping.

block (in this case 4). The resulting quotient is referred to as the logical block number (LBN), i.e., LBN $= 1$, and the remainder represents an offset (i.e., 1) within the block for the corresponding page. Since the mapping table only contains logical-to-physical block mapping, the size of the mapping table is $\lceil (log_2B)/8 \rceil \times B$ bytes, where B represents the number of blocks in the SSD. For example, a 1 TB SSD with 4 KB pages and 64 pages per block requires 2^{22} entries each with a 3-byte address. This leads to a mapping table size of 12 MB, which is a significant reduction compared to a table size of 1 GB for page-level mapping. However, a new (free) block has to be provided each time an update is made to the same logical page, and updating only part of a block causes expensive copy operations to move the valid pages from the old block to the new block. For example, update operations to pages 01, 05, 09, and 13 in Fig. 19 would each require a new block. Thus, the performance overhead due to "erase-before-write" is substantial because address mapping is performed at a block granularity [8, 24].

5.3 Hybrid mapping

In *hybrid mapping*, a logical block maps to a physical block using the block-level mapping scheme, and then the page-level mapping scheme is used to handle the updated pages within a block. Hybrid mapping maintains two types of blocks: data and log blocks [22]. *Data blocks* hold ordinary data and are managed using block-level mapping while *log blocks* are used for update requests and are managed using page-level mapping. For example,

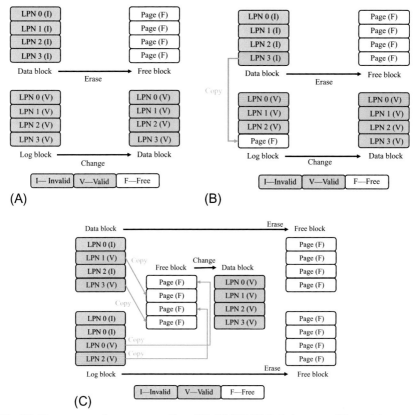

Fig. 20 Three types of merge operations [22, 25, 26]. (A) Switch merge; (B) partial merge; and (C) full merge.

if *write*(5, *E*0) is performed multiple times, the first write request is stored in the data block and then the subsequent update requests are written to the same log block. As another example, if the sequential write requests *write*(8, *A*0), *write*(9, *B*0), *write*(10, *C*0), and *write*(11, *D*0) are performed twice, the first set of sequential writes are written to the data block, while the second set of sequential writes are written to the same log block.

When all the log blocks are used up, *merge operations* have to be performed to secure new log blocks. Fig. 20 depicts three types of merge operations: switch merge, partial merge, and full merge. A *switch merge* is performed when update operations are performed sequentially to all the pages in a block [22, 25, 26]. This is illustrated in Fig. 20A, where update operations are performed on LPNs 0–3. After the update operations are completed, the original data block is erased (i.e., freed) and the log block becomes the new data block. The switch merge is the least costly operation because it only requires one erase operation. A *partial merge* shown in Fig. 20B is performed when a data block is partially updated, e.g., update operations are performed on LPNs 0–2. This requires additional copy operations to move the pages from the data block that were not updated to the corresponding entries in the log block. Then, the log block becomes the new data block and the old data block is erased. A *full merge* operation is performed when random writes occur, e.g., update operations are performed on LPNs 0, 0, 0, and 2 as shown in Fig. 20C. A full merge requires another new block to combine the valid pages from the original data block and the log block. The full merge is the most expensive among the three operations because it requires n copy operations (where n is equal to number of valid pages in a block) and two erase operations [25].

Hybrid mapping schemes take advantage of both page-level mapping and block-level mapping. There are several hybrid mapping schemes to manage log blocks, such as the Log-Block scheme [22], fully associativity sector translation (FAST) [24], the SuperBlock scheme [4], and locality aware sector translation (LAST) [25]. All these methods rely on log blocks; however, they differ on how the log blocks are handled. The following discusses these four techniques.

5.3.1 Log-Block

The seminal work on hybrid mapping is the *Log-Block scheme*, which maintains several log blocks in order to efficiently perform updates to the same logical pages [22]. This method is also referred to as the *block-level associativity sector translation* (BAST) scheme.

Fig. 21 illustrates the Log-Block scheme with four data blocks and three log blocks each with four pages. Suppose 12 consecutive write requests are made by the file system as shown. The first two requests *write*(5, *E*0) and *write*(6, *F*0) are mapped to Pages 01 and 02, respectively, in Block 00 using the block-level mapping scheme. However, the third and fourth requests *write*(5, *E*1) and *write*(5, *E*2) update Page 01 in Block 00, and thus a log block (Block 20) is allocated. Then, these two random write requests are written to Block 20, and the corresponding entry in the mapping table for the log block is updated. If LPN 5 is updated again later, these writes will be performed on Block 20 until it is fully filled up. In the same manner, the write requests *write*(8, *A*0), *write*(9, *B*0), *write*(10, *C*0), and *write*(11, *D*0) are written to Block 02. Afterward, the sequential write requests *write*(8, *A*1), *write*(9, *B*1), *write*(10, *C*1), and *write*(11, *D*1) update the Pages 08–11 in Block 02.

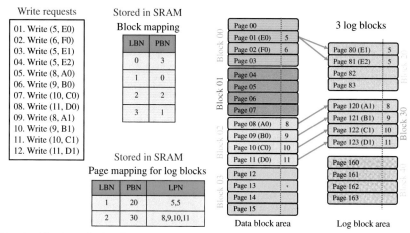

Fig. 21 The Log-Block scheme.

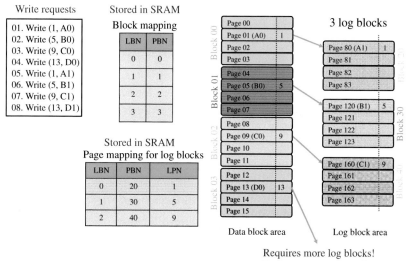

Fig. 22 Example of log block thrashing.

Thus, these requests are written to a new log block (Block 30) and the corresponding entry in the mapping table for the log block is updated.

In the Log–Block scheme, the cost of merge operations increases when frequent updates are made to the pages in different blocks [24]. For example, suppose write requests are issued to LPNs 1, 5, 9, 13, 1, 5, 9, and 13 as shown in Fig. 22. Due to *one-to-one associativity* between data and log blocks, the first sequence of write requests to LPNs 1, 5, 9, and 13 are written to different

Write requests

01. Write(0, A0) / 02. Write(1, B0) / 03. Write(3, D0) / 04. Write(2, C0)

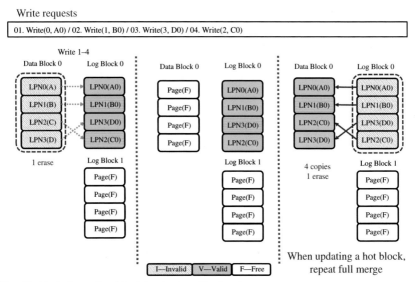

Fig. 23 Updating a hot block in the Log-Block scheme.

data blocks, and the second sequence of write requests for LPNs 1, 5, 9, and 13 will be written to different log blocks. However, all the log blocks will be used up by the time the second write request for LPN 13 is issued. Therefore, a merge operation has to be performed in order to secure a new log block. If write requests to LPNs 1, 5, 9, and 13 repeat, a merge operation is required for each write request leading to *log block thrashing* [24]. In addition, performance degradation occurs when one logical block is updated frequently with nonsequential write requests. For example, suppose write requests are made to LPNs 0, 1, 2, and 3, and then another write requests are made to LPNs 0, 1, 3, and 2 as shown in Fig. 23. Since the second sequence of write requests are not ordered, a full merge operation has to be performed to preserve the page ordering within a data block requiring four copy and one erase operations.

5.3.2 Fully associativity sector translation

The *fully associativity sector translation* (FAST) scheme utilizes *many-to-one associativity* to allow pages from multiple data blocks to be written to a single log block thereby circumventing the shortcoming of the Log-Block scheme [24]. FAST maintains two types of log blocks: one log block for sequential writes (*SW log block*) and a number of log blocks for random writes (*RW log blocks*). When an update request occurs, it is written to either the SW log block or a RW log block. An update request is written to the SW log block if either of

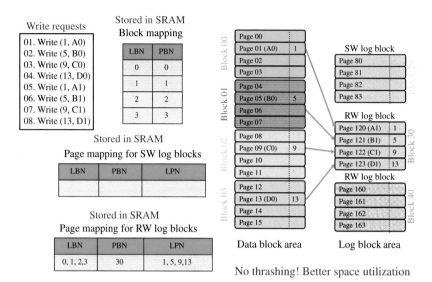

Fig. 24 The FAST scheme.

the following two conditions are satisfied: (1) the offset of the LPN being written to (i.e., LPN % 4) is 0 or (2) the write request is sequential with an offset that would fill the rest of the SW log block (i.e., 1, 2, or 3). Otherwise, it is written to a RW log block.

Fig. 24 illustrates the FAST scheme with write requests to LPNs 1, 5, 9, 13, 1, 5, 9, and 13. The offsets of the second sequence of writes to LPNs 1, 5, 9, and 13 are not zeros, thus they are all updated in a RW log block (Block 30). Therefore, unlike the Log-Block scheme, FAST does not suffer from log block thrashing since a merge operation is not required until all the pages in a RW log block are fully utilized [24, 26]. Furthermore, when non-sequential write requests update all the pages of a block, FAST performs a switch merge operation instead of an expensive full merge operation required by the Log-Block scheme. This is illustrated in Fig. 25, which shows how FAST handles write requests to LPNs 0, 1, 2, 3, 0, 1, 3, and 2. FAST maintains the sequential order of offsets within the SW log block, and as a result only a switch merge is needed when the SW log block is reclaimed [24].

The FAST scheme efficiently utilizes log blocks by separating sequential and random writes. The authors in [24] showed that the performance of FAST with 4–8 log blocks is similar to the Log-Block scheme with more than 30 log blocks. Moreover, FAST reduces the garbage collection overhead by 30% compared to the Log-Block scheme [25].

Write requests

01. Write(0, A0) / 02. Write(1, B0) / 03. Write(3, D0) / 04. Write(2, C0)

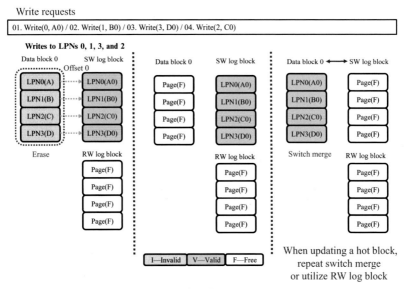

Fig. 25 Updating a hot block in the FAST scheme.

5.3.3 SuperBlock

The *SuperBlock* scheme increases the chances of performing switch or partial merge operations instead of expensive full merge operations to minimize performance degradation. The SuperBlock scheme combines N neighboring logical blocks into a *superblock* in order to exploit block-level spatial locality. The pages in a superblock can be freely mapped to any physical pages within the same superblock using a three-level page-mapping scheme [4]. The SuperBlock number and the page global directory (PGD) are stored in SRAM and used to map logical blocks. The page middle directory (PMD) points to the location of the page table (PT). Finally, the page table contains the physical block numbers and the physical page numbers. Both PMD and PT are stored in the spare area of the flash memory in order to reduce the overhead of SRAM. The following example shows how *SuperBlock* can increase the chance of performing switch or partial merge operations compared to the FAST scheme.

Fig. 26 shows switch and full merge operations in the SuperBlock scheme, where "hot" indicates pages that have been recently updated and are likely to be updated again in the near future, whereas "cold" indicates pages that have not been updated. When a switch merge operation is performed on a log block (Log Block 0) and a superblock consisting of two data

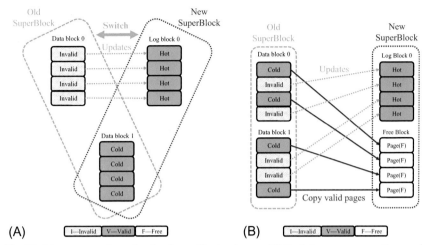

Fig. 26 Merge operations for the SuperBlock scheme. (A) Switch merge operation and (B) full merge operation.

blocks (Data Blocks 0 and 1), the log block becomes the new data block which is then combined with the valid data block (i.e., Data Block 1) into a new superblock. This is shown in Fig. 26A. When a full merge operation is performed as shown in Fig. 26B, the log block and a free block are first grouped into a new superblock. Then, all the valid pages from the old super-block are copied to the free block in the new superblock.

The updated (i.e., hot) pages are grouped into the same block (i.e., Log Block 0) to increase the chances of performing switch merge operations. This is because the hot pages in a data block will likely be updated to a new log block in the near future. When all the pages in a data block are inva-lidated due to updates, the new log block can be converted to a data block without copy operations. If hot and cold pages are intermixed in the same block, a full merge operation is required to copy valid (i.e., cold) pages.

In order to illustrate the advantage of the SuperBlock scheme, Fig. 27 compares the merge operations for FAST and SuperBlock when LPNs 0–7 have already been written to and update requests are issued to LPNs 1, 3, 5, 6, 1, 3, 5, and 6. For FAST shown in Fig. 27A, the first sequence of write requests are written to a RW log block. After the second sequence of write requests are performed, all the RW log blocks are used up and a full merge operation has to be performed to reclaim RW log blocks. The full merge operation generates two new data blocks: One for LPNs 1 and 3 from RW Log Block 1 with LPNs 0 and 2 from Data Block 0 and another for

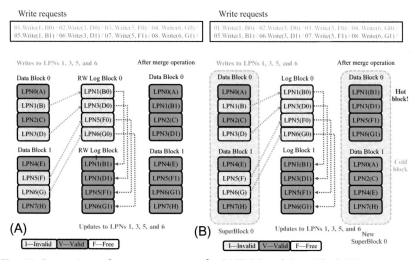

Fig. 27 Comparison of merge operations for FAST (A) and SuperBlock (B).

LPNs 5 and 6 from RW Log Block 1 with LPNs 4 and 7 from Data Block 1. If the same set of write requests repeats, FAST has to perform a full merge operation each time.

On the other hand, the SuperBlock scheme combines the two data blocks into a superblock as shown in Fig. 27B. The first sequence of write requests are written to a log block (Log Block 0). Then, the second sequence of write requests requires a full merge operation. However, the hot pages (LPNs 1, 3, 5, and 6) in the Data Block 0 and the cold pages (LPNs 0, 2, 4, and 7) in the Data Block 1 are separated by dynamically rearranging the pages as the new SuperBlock 0 shown in Fig. 27B. If the same sequence of write requests repeats, a switch merge instead of a full merge can be performed. Therefore, the SuperBlock scheme reduces the garbage collection overhead by reducing the frequency of full merge operations.

The authors in [4] showed that the garbage collection overhead is reduced by 23% when the superblock size N is 32 compared to when there is no superblock (i.e., N is one). They also show that the SuperBlock scheme (with N equal to 32) decreases the garbage collection overhead by 35% compared to FAST [4]. However, the read performance of the SuperBlock scheme can be reduced due to the three-level page mapping scheme because PMD and PT need to be read from the spare area of the flash memory to generate the mapping table.

5.3.4 Locality-aware sector translation

The *locality-aware sector translation* (LAST) scheme was proposed in order to consider both spatial and temporal locality of access patterns [25]. This scheme exploits spatial locality by using SW log blocks (similar to FAST) and temporal locality by distinguishing blocks as hot and cold (similar to SuperBlock).

Fig. 28 shows the architecture of the LAST scheme. The *Locality Detector* distinguishes between random and sequential write requests. If the length of sequential write requests is longer than a threshold (i.e., 8), they are written to SW log blocks; otherwise, they are written to RW log blocks. Therefore, multiple SW log blocks are used to log long sequential writes to exploit spatial locality, while RW log blocks are used to exploit temporal locality by distinguishing hot and cold regions to avoid expensive full merge operations.

For random writes, data are distinguished into hot and cold *partitions* based on how often they are updated. The hotness of each page is determined based on its *update interval*, which is determined by the distance between updates to the same logical page. For example, if writes are performed on LPNs 0, 3, 2, and 0, the update interval of LPN 0 is 2 because there are two other requests (LPNs 3 and 2) between the first and second writes to LPN 0. If the update interval of a page is less than k as a criterion, it is considered as a hot page; otherwise, it is treated as a cold page. Fig. 29

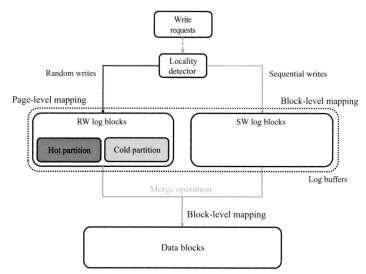

Fig. 28 The architecture of LAST [25].

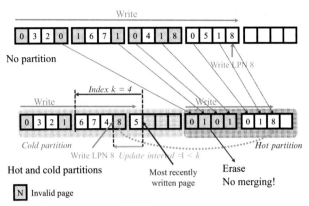

Fig. 29 Hot and cold partitions for LAST [25].

compares a scheme with no partitioning with a scheme that partitions pages into hot and cold. When partitioning is not used, the requests are sequentially written to the random log blocks regardless of whether pages are hot or cold. On the other hand, when hot and cold pages are partitioned, the second write to LPN 0 is performed on the hot partition because its update interval is 2, which is less than k (i.e., $k = 4$). Similarly, writes to LPNs 1 and 8 are also considered as hot [25].

LAST exploits both spatial and temporal locality by using SW log blocks and hot/cold partitions. However, it cannot efficiently detect a workload consisting of a series of short sequential writes instead a series of long sequential writes [26]. For example, if the length of sequential writes is shorter than the threshold, they are not stored in SW log blocks. Furthermore, the threshold should be carefully selected. If the threshold is too long, many sequential writes will be treated as random writes. If the threshold is too short, many random writes will be handled using the SW log blocks. The authors in [25] show that for certain applications LAST reduces garbage collection overhead by 86%, 81%, and 46% compared with the Log-Block scheme, FAST, and SuperBlock, respectively.

5.3.5 Demand-based flash translation layer

The *Demand-based flash translation layer* (DFTL) scheme was proposed to reduce the mapping table size in the SRAM for the page-level mapping scheme [26]. The basic idea is to store the page-based mapping table in the flash memory and then cache the entries on demand in the *cached mapping table* (CMT) in the SRAM. *Translation blocks* in the flash memory hold the entire page-level mapping table, and the *global translation directory* (GTD) in the SRAM tracks these translation blocks. Storing the mapping table in the flash memory causes performance overhead for read requests because the translation blocks need to be accessed first from the flash memory to bring the CMT entries into the SRAM. Therefore, the DFTL stores small GTD and CMT in the SRAM instead of a large page-based mapping table. The DFTL improves response time by 78% compared to FAST [26]. However, storing the mapping table in the translation blocks causes a serious performance overhead for random read requests even with CMT. Furthermore, extra write operations are required to update the translation blocks.

6. Garbage collection

Erase operations are performed at a block granularity due to their long latency and the limitations of flash devices. When write requests are issued for logical pages, they have to be written to free pages (i.e., out-of-place updating), and the original pages are marked as invalid. As a result, many invalid pages are generated and these obsolete pages cannot be overwritten due to the characteristics of flash memories as discussed in Section 2. Therefore, the FTL has to prepare and maintain a pool of free blocks for write operations. *Garbage collection* (GC) is the reclamation process that invalidates obsolete blocks and make them available for future write operations.

The main issues for garbage collection are minimizing its cost and maximizing the reclaimed space, and the important metrics used to evaluate different garbage collection schemes are the write amplification factor and the garbage collection efficiency. The *write amplification factor* is the ratio of the actual number of written pages and the number of page writes [23]. On the other hand, the *garbage collection efficiency* is defined as the average number of invalid pages in each victim block to be erased [27]. An optimized garbage collection algorithm reduces write amplification, thus improving the performance and lifetime of flash devices. For example, Fig. 30 illustrates the importance of *victim selection* in garbage collection [28]. In Scenario 1, Blocks 1 and 2 are chosen as victim blocks, and the valid pages in these two blocks are moved to Block 7. Afterward, Blocks 1 and 2 are erased to make them free blocks. On the other hand, in Scenario 2, Blocks 1 and 6 are chosen as victim blocks, which do not have any valid pages, and thus

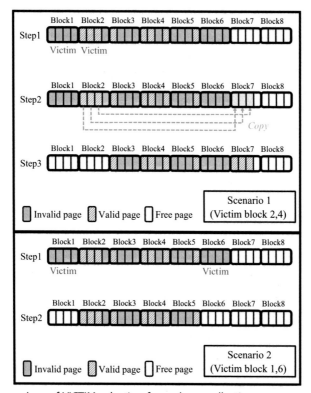

Fig. 30 Comparison of VICTIM selection for garbage collection.

no data needs to be moved. As can be seen, even though both Scenario 1 and 2 have the same data, Scenario 2 leads to a more efficient garbage collection process than Scenario 1 due to better victim selection. Existing garbage collection algorithms differ on how victim section is performed, which include first in first out (FIFO), Greedy, Windowed, and d-choice [29]. Fig. 31 illustrates these algorithms, and Sections 6.1–6.4 discuss each of these techniques.

6.1 FIFO GC algorithm

The *FIFO GC algorithm* chooses blocks to be reclaimed in a cyclic manner where the first written block is erased first [29]. For example, in Fig. 31, the FIFO GC algorithm will reclaim Blocks 1 through 6 sequentially, even though over 50% of the pages in Blocks 1 and 2 are still valid. Therefore, the FIFO GC algorithm is easy to implement, but it wastes many valid pages and increases write amplification [29].

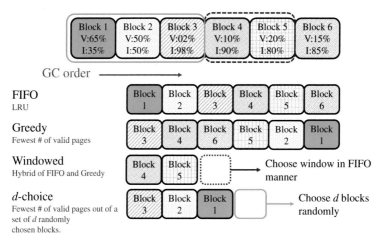

Fig. 31 Garbage collection algorithms.

6.2 Greedy GC algorithm

In contrast to the FIFO GC algorithm, the *Greedy GC algorithm* selects blocks that contains the fewest number of valid pages [30]. As shown in Fig. 31, the Greedy GC algorithm will select Blocks 3, 4, 6, 5, 2, and 1 in order representing the fewest number of valid pages, which allows Blocks 1 and 2 to be used longer until they are reclaimed. Hence, the Greedy GC algorithm results in the least amount of write amplification during random write operations [30]. However, the implementation overhead of the Greedy algorithm is significantly higher than the FIFO algorithm.

6.3 Windowed GC algorithm

The *Windowed GC algorithm* take advantages of both FIFO and Greedy algorithms [29]. The algorithm selects a window containing a list of blocks using the FIFO algorithm. From this selected window, the algorithm then chooses the block that has the fewest number of valid pages using the Greedy algorithm. Fig. 31 illustrates the Windowed GC algorithm with a window size of two blocks. In this example, Blocks 4 and 5 are first chosen using the FIFO algorithm. Then, Blocks 4 and 5 are selected sequentially using the Greedy algorithm. The most appropriate window size is a design trade-off between simplicity of implementation and resulting write amplification. As the window size increases, the complexity increases. On the other hand, as the window size decreases, the amount of write amplification increases [29].

6.4 *d*-Choice GC algorithm

The *d-choice GC algorithm* is similar to the Windowed GC algorithm, but the set of *d* blocks is chosen randomly instead of in a FIFO manner [29]. From this chosen *d* blocks, the algorithm reorders the blocks for garbage collection using the Greedy algorithm. For example, in Fig. 31, Blocks 1, 2, and 3 are selected randomly, and then they are reordered based on the fewest invalid pages. As a result, the *d*-choice GC algorithm selects Blocks 3, 2, and 1 in sequence [29].

7. Wear leveling

As discussed in Section 2, flash devices have a finite number of P/E cycles as a reliable memory. If a subset of blocks are used excessively, then these blocks would wear out sooner and, as a result, the total storage capacity will decrease over time. This decrease in capacity seriously impacts the performance because if there are not enough free blocks to support write requests, garbage collection needs to be performed more frequently to secure additional free blocks degrading performance [27]. Therefore, uneven wearing of flash blocks leads to not only capacity degradation but also performance degradation. To combat this problem, wear-leveling algorithms try to wear out flash blocks more evenly. *Dynamic wear leveling* techniques select the blocks with the lowest number of P/E cycles when new write requests are issued [1, 31]. However, dynamic wear leveling operates only on the currently active blocks, and does not consider the blocks that have not been updated for a long time, such as the ones that contain the operating system. Thus, *static wear leveling* techniques changes the state of inactive blocks to active blocks [1].

Fig. 32 illustrates the reason why both static and dynamic wear leveling are required to avoid worn out flash blocks. The blocks that are hot are frequently updated, while the blocks that are cold are rarely updated. In this example, the P/E cycle counts for the cold block and the hot block are 1 k and 5 k, respectively. If the dynamic wear leveling technique is used exclusively for 5 k P/E cycles, only the hot block will be worn out because the cold block has not been not updated as shown on the left side of Fig. 32. In contrast, if both static and dynamic wear leveling are used, the P/E cycle limit of 10 k (in case of SLC) can be avoided by swapping the hot block with the cold block and distributing the number of P/E cycles between the two blocks as shown on the right side of Fig. 32.

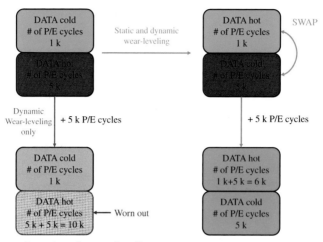

Fig. 32 Dynamic and static wear leveling.

Numerous wear leveling techniques have been proposed in the literature [5, 31–33]. Sections 7.1 and 7.2 discuss the Evenness-Aware and the dual-pool algorithms. The concept behind these two algorithms are similar, but how they are triggered is different.

7.1 Evenness-aware algorithm

The *Evenness-Aware algorithm* was proposed to improve the endurance using the block erasing table (BET) that maintains information about the number of P/E cycles within a certain period of time [5]. The algorithm minimizes the difference in P/E cycles between any two blocks using a cyclic-queue-based scanning procedure to evenly distribute P/E cycles among the blocks. A software leveler is triggered whenever a block is recycled by the garbage collection process. The algorithm then performs static wear-leveling based on a preset condition, which is the difference between the maximum and the minimum number of P/E cycles. If this difference is larger than a preset condition, garbage collection with static wear leveling is performed and information in the BET is updated. Therefore, Evenness-Aware algorithm avoids static data remaining at any block for a long time [5].

7.2 Dual-pool algorithm

The *dual-pool algorithm* maintains two types of data pools: cold pool and hot pool [31]. The *cold pool* contains blocks that are updated infrequently, such as memory blocks that store the operating system, while the *hot pool* contains blocks that are frequently updated. The P/E cycle counts increase slowly for the blocks in the cold pool, while the P/E cycle counts increase quickly for the blocks in the hot pool due to frequent updating. The main idea of the dual-pool algorithm is to monitor the P/E cycle count of each block, and then swap a block from the hot pool with a block from the cold pool when the P/E cycles of the hot block is larger than that of the cold block by a predefined threshold [31].

8. Bad block management

Flash blocks can become bad due to faults during production process or become worn out after a limited number P/E cycles. Bad block management avoids these bad blocks to maintain high reliability [6, 7]. Sections 8.1 and 8.2 discuss the process of bad block management, which consists of *bad block recognition* and *bad block replacement*.

8.1 Bad block recognition

There are a couple of ways to detect bad blocks: Using the code "FFh" and error correction code (ECC) [7]. "FFh" is a special code included with each valid block (located at either 1st or 2nd page of a valid block). Note that "FFh" basically represents a byte that has been erased (i.e., all the bits represent data "1"). Therefore, if this "FFh" is missing, it means that an erase failure occurred.

8.2 Bad block replacement

The FTL creates a table of bad blocks from the bad block recognition. A bad block replacement is required when the target address of a request matches with the address of a bad block. There are two types of bad block replacement schemes: Skip Block and Reserve Block methods. The *Skip Block* method avoids the use of a bad block by writing the data to the block right after the bad block [7]. The Skip Block method is depicted in Fig. 33. In this example, the 2nd block in the device is bad; therefore, this block is skipped and the 3rd block is chosen instead. In the *Reserved Block* method, a good block from the reserved block area (RBA) is chosen to write the data when a bad block is recognized [7]. The Reserve Block method is illustrated in Fig. 34. In this example, the 2nd block is detected to be a bad block as in the previous example, thus the data 2 is written to a good block (referred to as a reservoir block) from the reserved block area.

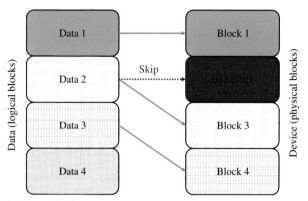

Fig. 33 Skip Block method.

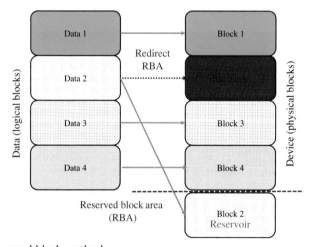

Fig. 34 Reserved block method.

9. SSD parallelism techniques

An SSD exploits parallelism within the flash memory to increase the overall bandwidth and amortizes the cost of long latency operations. This is achieved by employing techniques such as *channel striping*, *flash-chip pipelining*, *die interleaving*, and *plane sharing* in order to achieve high performance [21, 34, 35]. Sections 9.1–9.5 discuss these methods in detail.

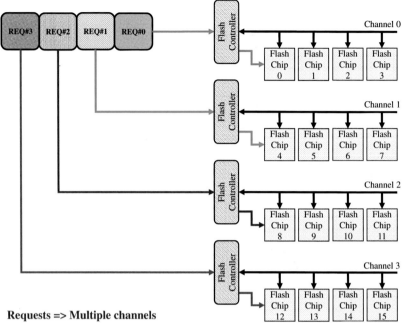

Requests => Multiple channels

Fig. 35 SSD channel striping.

9.1 Channel striping

The *Channel Striping* technique exploits channel-level parallelism in a multi-channel architecture. This is illustrated in Fig. 35, where multiple-requests are distributed over the multiple channels in a *round-robin* fashion. For example, the four write requests are allocated to Channels 0–3 as shown in Fig. 35 [2, 3, 21, 35].

9.2 Flash-chip pipelining

The *Flash-chip pipelining* technique exploits flash-chip-level parallelism by utilizing multiple flash-chips on a channel as shown in Fig. 36. In this example, the four write requests are assigned to the four flash-chips on Channel 1 [35].

9.3 Die interleaving

The *die interleaving* technique exploits die-level parallelism using the *interleaved program page* command. This is depicted in Fig. 37, where Request 0 and Request 1 are allocated to Die 0 and Die 1, respectively, on the same flash-chip (i.e., Flash-chip 0)[16, 35].

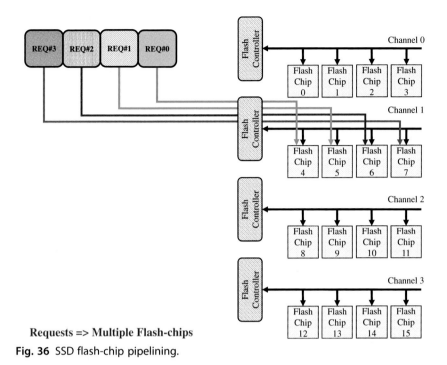

Requests => Multiple Flash-chips

Fig. 36 SSD flash-chip pipelining.

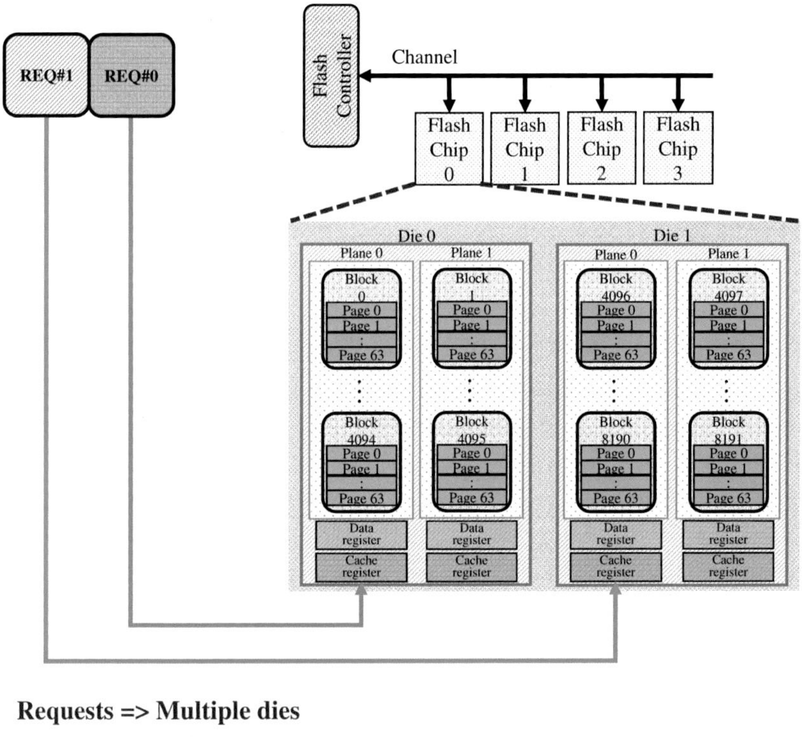

Requests => Multiple dies

Fig. 37 SSD die interleaving.

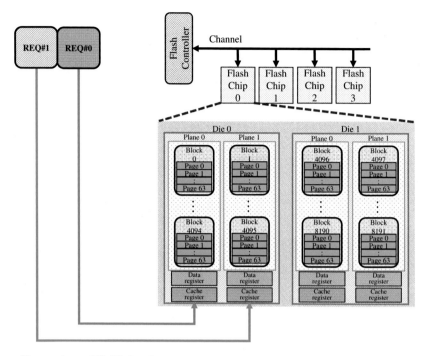

Requests => Multiple planes

Fig. 38 SSD plane sharing.

9.4 Plane sharing

The *plane sharing* technique exploits plane-level parallelism using the *Two-plane Page Program* command. This is shown in Fig. 38, where Request 0 and Request 1 are assigned to Plane 0 and Plane 1, respectively, on Die 0 in Flash-chip 0 [16, 35].

9.5 Combining parallelism techniques

All three techniques (*channel striping*, *flash-chip pipelining*, and *die interleaving* (or *plane sharing*)) can be combined together to exploit multiple levels of parallelism. This is illustrated in Fig. 39, where there are 8 requests. The Requests 0–3 and 4–7 are assigned to different channels according to the *channel striping* technique. The requests 0–1 (or 4–5) and 2–3 (or 6–7) are allocated to different flash-chips using the *flash-chip pipelining* technique. Finally, Requests 0 and 1 (or 2 and 3, 4 and 5, 6 and 7) are assigned to different dies using the *die interleaving* technique. Therefore, combining these parallelism techniques can provide significant improvement in throughput

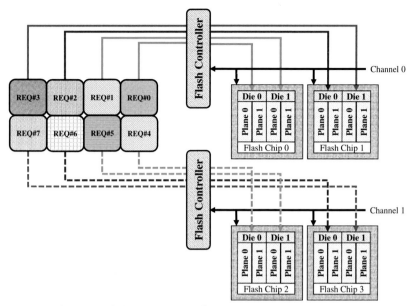

Fig. 39 Combination of channel striping, flash-chip pipelining, and die interleaving.

and performance [21]. In addition, the order of how these parallelism techniques are applied is important for the performance of SSDs. This will be discussed in the following section.

10. Page allocation

As mentioned in Section 9, write requests can be distributed among multiple channels, chips, dies, and planes using channel striping, flash-chip pipelining, die interleaving, and plane sharing, respectively. The distribution of write requests requires the allocation of pages, and thus the order of page allocation among channels, flash-chips, dies, and planes is important for improving the performance of not only writes but also subsequent reads and updates [35, 36].

Static allocation assigns logical pages to free physical pages in the flash memory based on a predetermined order of channel, flash-chip, die, and plane. On the other hand, *dynamic allocation* considers P/E cycles of blocks, idle/busy state of flash-chips, and/or idle/busy state of channels. Studies have shown that static allocation schemes consistently outperform dynamic allocation schemes for read operations. In contrast, dynamic allocation schemes are designed to improve the lifetime of SSDs [36].

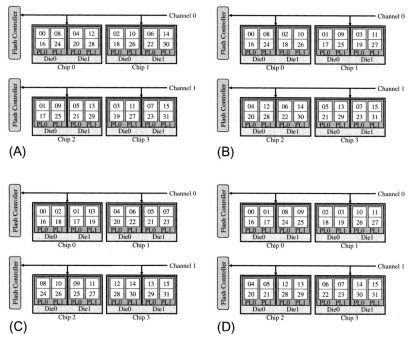

Fig. 40 Page allocation strategies. (A) Channel → Flash-chip → Die → Plane. (B) Flash-chip → Die → Channel → Plane. (C) Die → Plane → Flash-chip → Channel. (D) Plane → Flash-chip → Channel → Die.

The performance of static allocation schemes can be evaluated based on a predetermined priority [35]. There are 24 possible static allocation schemes because the permutation among channel, flash-chip, die, and plane is 4! The *channel-first* allocation exploits channel-level parallelism using *channel striping*, while the *flash-first* allocation maximizes the benefit of flash-chip-level parallelism using *flash-chip pipelining*. The *die-first* allocation takes advantage of die-level parallelism using *die interleaving*. The *plane-first* allocation scheme exploits plane-level parallelism using *plane sharing*, where flash operations are activated concurrently on multiple planes [35].

The best allocation schemes for channel-first, flash-first, die-first, and plane-first are shown in Fig. 40A–D, respectively. The arrows indicate the order of how the resources are allocated. For example, "Channel → Flash-chip → Die → Plane" (CFDP) indicates that the request is stripped in order across multiple channels, then across multiple flash-chips, then across multiple dies, and finally across multiple planes as shown in Fig. 40A. The channel-first and the flash-first allocation policies are better for latency sensitive applications, while the die-first and the plane-first allocation policies are better for throughput sensitive applications. Among these page allocation strategies, the optimal page allocation strategy is the Plane $\rightarrow$ Flash-chip $\rightarrow$ Channel $\rightarrow$ Die (PFCD) allocation scheme [35].

11. Recent research trend

Performance, lifetime, and reliability are all important factors for SSDs, and tradeoffs exist among these factors. This section focuses on recent research efforts on improving these factors. Table 1 provides a taxonomy of various research efforts on FTL functionalities as well as non-FTL-related approaches to improve performance, lifetime, and reliability. Since lifetime and reliability of SSDs are highly related, they are combined into the same category.

11.1 Performance

Performance is the most important metric for SSDs in enterprise applications. Therefore, most FTL schemes focus on boosting performance of SSDs. Currently, garbage collection is the main culprit for performance degradation. Sections 11.1.1–11.1.10 present recent techniques that focus on performance.

11.1.1 Host Interface I/O Scheduler

The *Host Interface I/O Scheduler* (HIOS) is a garbage collection and quality of service (QoS) aware FTL [21]. This is achieved by redistributing the garbage collection overhead across noncritical I/O requests as well as performing *Channel QoS-aware scheduling*. The authors in [21] observed that the worst-case latency of SSDs due to garbage collection and channel contention can vary between 210 and 4670 ms.

As shown in Fig. 41, HIOS classifies I/O activities based on their current level of *deadline satisfaction*. If an I/O request takes longer than the require deadline, it is defined as a *negative activity* (i.e., critical I/O request). If an I/O request takes less than but close to the required deadline, it is categorized as a *neutral activity*. If an I/O request takes much less than the required deadline, it is classified as a *positive activity* (i.e., noncritical I/O request). The main idea of HIOS is to predict which I/O requests are negative activities and distributes their overhead across positive activities.

In order to predict whether an I/O request is a negative activity, HIOS analyzes the length of the I/O request (i.e., the number of writes) and its

Table 1 Taxonomy of recent research trends.

FTL operations	Factors	
	Performance	Lifetime and reliability
Address mapping and allocation	– Real-time flash translation layer (RFTL) [37]	
	– Multilevel associated sector translation (MAST) [38]	
	– Zone-based flash translation layer (Z-MAP) [39]	
	– Convertible flash translation layer (CFTL) [40]	
Garbage collection	– Host Interface I/O Scheduler (HIOS) [21]	– Stable Greedy garbage collection [41]
	– Real-time flash translation layer (RFTL) [37]	– Garbage collection aware striping (GCAS) [42]
	– Lazy real time garbage collection (Lazy-RTGC) [43]	
	– Stable Greedy garbage collection [41]	
	– Garbage collection aware striping (GCAS) [42]	
Wear leveling		– BER-based wear leveling and bad block management [33]
		– Zombie NAND [17]
Bad block management		– Zombie NAND [17]
Others	– Willow [44]	– Self-healing SSDs [45]
	– An Effective Page Padding method for RAM buffer algorithms [46]	– Dynamic program and erase scaling (DPES) [47]
	– Harey Tortoise [48]	– An Effective Page Padding method for RAM buffer algorithms [46]
		– Flash correct-and-refresh (FCR) [49]
		– Retention optimized reading (ROR) and retention failure recovery (RFR) [50]

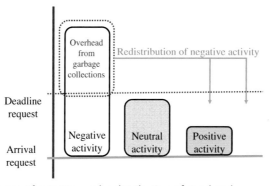

Fig. 41 Three types of activities and redistribution of overhead.

LPNs, then determines the target physical block(s) to be written to. Moreover, if the number of free pages available in the target physical block(s) is less than the length of the I/O request, additional blocks need to be freed in order to satisfy the request. Therefore, the garbage collection time is also estimated. Based on this, HIOS predicts whether or not this I/O request is a negative activity. If it is predicted as a negative activity, its garbage collection overhead is scheduled across the available time slots of positive activities.

Channel QoS-aware scheduling in HIOS reorders I/O requests based on *Empty Channel First*. This is illustrated in Fig. 42, which compares FIFO I/O scheduling and Channel QoS-aware scheduling. Suppose Requests 0–5, which are neutral activities, are issued to Channels 0, 1, 0, 2, 0, and 3, respectively. Fig. 42A illustrates the FIFO I/O scheduling, where Request 2 has to wait until Request 0 is completed due to channel contention, and this delay affects Requests 3–5. On the other hand, the Channel QoS-aware scheduling shown in Fig. 42B removes channel contention by interchanging Requests 2 and 3 because Channel 2 is empty. HIOS has shown to reduce worst-case latencies by 86.6% compared to FIFO I/O scheduling used in commercial systems [51].

11.1.2 Real-time flash translation layer
The *real-time flash translation layer* (RFTL) scheme considers real-time service guarantees by focusing on reducing worst-case response time [37]. This is achieved by employing a distributed partial garbage collection scheme with the real-time task scheduler and a real-time aware hybrid addressing mapping technique.

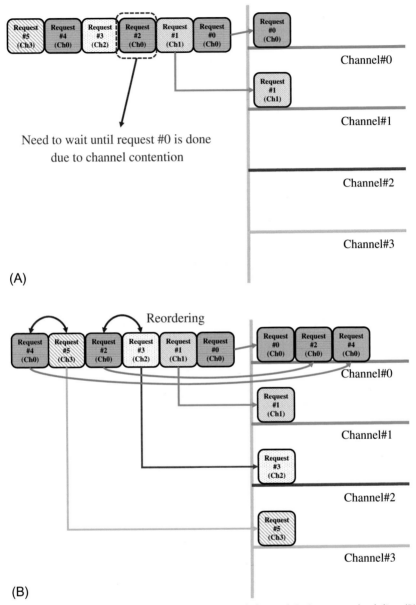

Fig. 42 Comparison of FIFO I/O scheduling (A) and channel QoS-aware scheduling (B).

As the name suggests, the *distributed partial garbage collection* scheme divides the garbage collection process using a time–cost model that considers the latencies of read, write, and erase operations as well as the number of valid pages that need to be copied from the victim block(s) to a new block. For

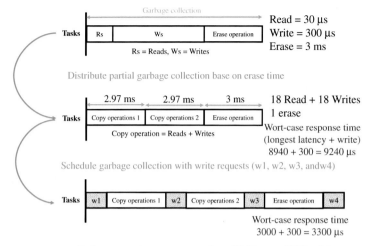

Fig. 43 Main concept of RFTL.

example, Fig. 43 illustrates the garbage collection process, which can be divided (i.e., distributed) into copy and erase operations. Suppose the latencies of read, write, and erase operations are 30 μs, 300 μs, and 3 ms, respectively, and the victim block has 18 valid pages that have to copied to a free block before the erase operation resulting in a total garbage collection time of 8940 μs (i.e., 5940 μs for read and write operations and 3000 μs for the erase operation). This process can be divided into copy operation 1 (9 read and 9 write operations), copy operation 2 (9 read and 9 write operations), and one erase operation requiring 2970 μs, 2970 μs, and 3000 μs, respectively. Afterward, the *real-time task scheduler* interleaves the distributed partial garbage collection workload with other pending tasks (i.e., write operations w1, w2, w3, and w4) to reduce the worst-case response time to 3300 μs as shown in Fig. 43.

Interleaving write requests with a distributed partial garbage collection can cause a coherence problem. This situation is explained using the example shown in Fig. 43. Suppose the write request w2 is to a valid page in the victim block, which has already been copied to a new block during copy operation 1. This causes a synchronization problem where the page that has been copied is no longer valid and the write request w2 must be properly reflected.

In order to handle this situation, RFTL utilizes the *real-time aware hybrid address mapping*, which maintains three types of physical block—primary, replacement, and buffer blocks as shown in Fig. 44. Initial write requests

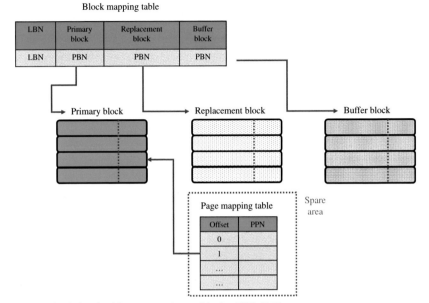

Fig. 44 The hybrid address mapping table for RFTL [37].

are first served by a *primary block*, which is similar to a data block in existing hybrid mapping schemes. The update requests are performed to a *replacement block*, which is similar to a log block in existing hybrid mapping schemes. When a garbage collection is triggered for the primary block, all of its valid pages are migrated to the corresponding *replacement block* during copy operations 1 and 2 shown in Fig. 43. Meanwhile, the pending write requests to this logical block (i.e., w2 and w3 in Fig. 43) are allocated to the corresponding *buffer block*. Afterward, the replacement block becomes the new primary block, and the freed primary block becomes the new replacement block. When the buffer block becomes full, then it becomes the new replacement block and the freed primary block becomes the new buffer block. These three blocks periodically change their roles in order to provide a write space during the distributed garbage collection process. A page-level mapping table is required to freely update pages among the three types of blocks. The page-level mapping table indicates which physical page belongs to one of the three types of physical blocks, and is stored in the spare area to reduce overhead in SRAM.

The RFTL scheme reduces worst-case latency of garbage collection. However, RFTL requires extra space to manage the three types of blocks. Furthermore, storing page-level mapping table in the spare area may effect latency of read operations.

11.1.3 Lazy real time garbage collection

The *lazy real time garbage collection* (Lazy-RTGC) scheme defers garbage collection as long as possible to optimize both the average and the worst-case system response time. This is achieved using on-demand page-level address mapping and partial garbage collection [43]. *On-demand page-level address mapping* stores the entire page-level address mapping table in the *translation blocks* and then the table entries are selectively cached in SRAM to provide the advantage of page-level mapping with a minimum SRAM cost [26]. On the other hand, *partial garbage collection* is similar to the distributed partial garbage collection scheme in RFTL (see Section 11.1.2), but the address mapping scheme and the scheduling method are different.

Unlike RFTL that uses hybrid mapping, Lazy-RTGC can handle update requests during a partial garbage collection because page-level mapping scheme is used. Moreover, it does not need extra space compared to RFTL. For these reasons, the authors in [43] show that the number of erase operations is reduced by 67.4% compared to RFTL and the SRAM overhead is reduced by 90.0% compared to pure page-level mapping.

However, read operations of Lazy-RTGC incur longer latency because the mapping table is stored in the translation blocks need to be read from flash memory. Furthermore, the mapping table in the translation blocks needs to be updated with every write request, which causes frequent garbage collection operations.

11.1.4 Stable Greedy GC

The Greedy GC scheme discussed in Section 6 is highly complex and scales poorly. Furthermore, it does not consider whether or not a block is hot during the victim block selection. In contrast, the *Stable Greedy GC* scheme takes into account hot and cold blocks during the victim block selection process [41]. This is achieved by considering *page lifetime*, which is the time between when new data is written to a page and when the page is invalidated due to an update. A page is considered to be hot if its lifetime is short, which is similar to the update interval used in LAST (see Section 5.3.4) [25]. Another important factor used in the Stable Greedy GC scheme is *Block Dormant Period*, which is the interval between the current time and when a page in a block was last invalidated. A short Block Dormant Period for a block indicates that the pages in this block are likely to be updated in the near future. The Stable Greedy GC scheme avoids selecting a victim block with a short Block Dormant Period because it is considered as an unstable block. Therefore, deferring the choice of these unstable blocks as victim blocks will reduce the number of copy operations during garbage collection.

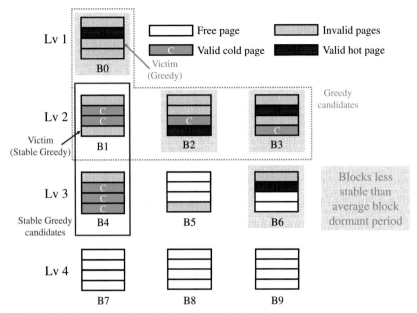

Fig. 45 Victim block selection for Greedy and Stable Greedy.

Fig. 45 compares victim block selection for Greedy and Stable Greedy, where Lv indicates the level of invalid pages, i.e., the blocks in Lv 1 have the most number of invalid pages followed by Lv 2 and so on. The Greedy scheme first chooses block B0 as the victim block because it has the most number of invalidated pages. Then, blocks B1, B2, and B3 are chosen as candidates due to the number of invalid pages these blocks have, and so on. In contrast, the Stable Greedy scheme avoids blocks B0, B2, B3, and B6 (highlighted by green regions) because their Block Dormant Periods are less than average. Therefore, blocks B1 and B4 are chosen as candidates, and then block B1 is selected as the victim block because it has more invalid pages than block B4. Therefore, Stable Greedy is an efficient and adaptive garbage collection scheme that has shown to outperform previous garbage collection algorithms under various workloads [41].

11.1.5 Multilevel associated sector translation

Previous hybrid mapping techniques such as the Log-Block scheme, FAST, LAST, and SuperBlock discussed in Section 5 do not consider workload characteristics. In contrast, *multilevel associated sector translation* (MAST) is a workload-aware address mapping scheme that dynamically detects hot and cold pages in order to avoid unnecessary merge operations [38].

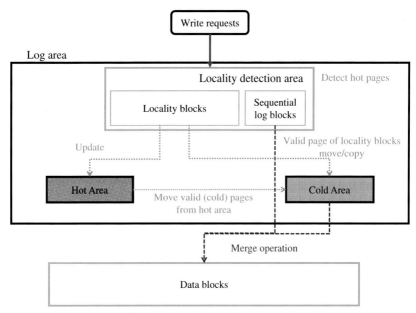

Fig. 46 The architecture of MAST.

Fig. 46 shows the architecture of MAST, which has three types of log areas—Locality Detection Area, Hot Area, and Cold Area. The *Locality Detection Area* has a small number of log blocks consisting of Locality Blocks and Sequential Log Blocks. The Locality Detection Area is used to distinguish between hot and cold pages for write requests and is similar to LAST (see Section 5.3.4), and Locality Blocks and Sequential Log Blocks follow the basic functionalities of RW and SW log blocks in FAST, respectively [24].

Initial sequential write requests in MAST are allocated to the Sequential Log Blocks. On the other hand, initial random write requests are allocated to the Locality Blocks in order to distinguish between hot and cold pages. When update requests are issued to pages in the Locality Blocks, they are considered as hot pages. Then, these update requests are written to the Hot Area.

The Cold Area receives cold pages from Locality Blocks and the Hot Area. When all the blocks in the Locality Detection Area are used up, all the valid pages are considered as cold pages and, as a result, they are copied to the Cold Area. If there is no more space in the Hot Area, its valid pages are no longer considered as hot and are moved to the Cold Area.

When Sequential Log Blocks in the Locality Detection Area and log blocks in the Cold Area are all used up, a merge operation is performed. Therefore, MAST reduces unnecessary copy operations by storing stable data such as sequential and cold data in data blocks, while unstable data such as hot data are stored in log blocks. As a result, it provides better performance in PC application workloads compared to FAST and LAST by 7% and 3%, respectively.

11.1.6 Dynamic hybrid address mapping techniques

Hybrid mapping techniques discussed in Section 5.3, such as the Log-Block scheme, FAST, LAST, SuperBlock, and MAST, are considered to be *static* schemes. Recently, *dynamic* hybrid mapping techniques that switch between page-level and block-level mappings have been proposed. These include *zone-based flash translation layer* (Z-MAP) [39] and *convertible flash translation layer* (CFTL) [40].

The Z-MAP scheme utilizes two types of zones: Page-mapping Zone and Block-mapping Zone. This is shown in Fig. 47. The *Page-mapping Zone* is used to store random data to better manage a large number of partial updates. On the other hand, the *Block-mapping Zone* is used store sequential data to reduce the size of the mapping table. These zones are dynamically allocated based on the workload. The Z-map scheme avoids expensive merge operations by classifying the workload before it is stored into flash

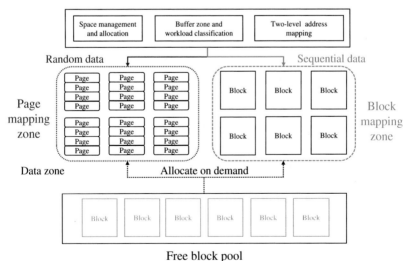

Fig. 47 The architecture of ZMAP.

memory, thereby improving the garbage collection efficiency. Therefore, Z-MAP is a workload adaptive technique that not only improves performance by reducing the garbage collection overhead but also reduces the mapping table size by using the Block-mapping Zone [39]. The authors in [39] show that Z-Map improved performance by 76%, the mapping table size is reduced by 81%, and the garbage collection overhead is reduced by 88% compared to FAST [24] and DFTL [26].

The CFTL scheme also dynamically switches between block-level mapping and page-level mapping based on data access patterns, i.e., block-level mapping is selected for read intensive data, while page-level mapping is adopted for write intensive data. In order to reduce the size of the page-level mapping table for write intensive workloads, CFTL employs a spatial locality-aware caching mechanism and adaptive cache partitioning [40]. The *spatial locality-aware caching mechanism* stores the mapping information only for the first page of a block of consecutive pages. For example, suppose LPNs 0, 1, 2, and 3 are mapped to PPNs 110, 111, 112, and 113, respectively. Then, the mapping information is maintained only for LPN 0 and PPN 110 and the number of consecutive pages (i.e., 4) as shown in Fig. 48. The *adaptive cache partitioning* is used to adjust the size of page-level and block-level mapping tables according to the workload. If the access pattern contains mainly write requests, the size of the page-level mapping table is increased. On the other hand, when the access pattern contains mainly read requests, the size of the page-level mapping table is decreased. As a result, CFTL outperforms DFTL [26] by up to 24% and 4% for read intensive workloads and write intensive workloads, respectively [40].

11.1.7 Write buffer management schemes

As discussed in Section 5, the mapping table is stored in the SRAM to minimize the address translation overhead. There are also techniques that use the

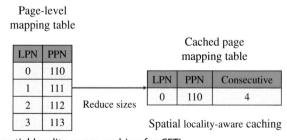

Fig. 48 The spatial locality-aware caching for CFTL.

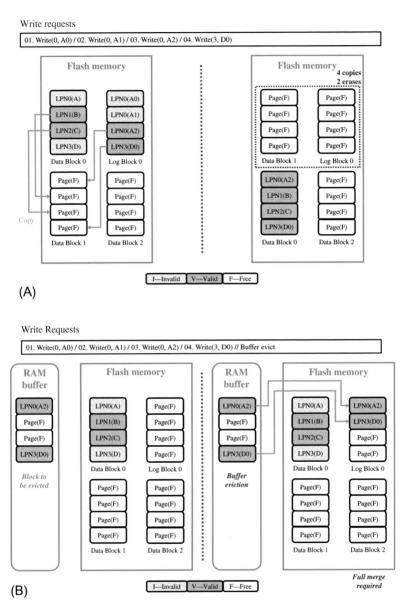

Fig. 49 Comparison without and with write buffers. (A) Write operations without a write buffer; and (B) write operations with a write buffer.

SRAM as a *write buffer* to increase the performance of random writes [52]. In order to illustrate the importance of a write buffer, suppose update requests are issued to LPNs 0 (A0), 0 (A1), 0 (A2), and 3 (D0) after LPNs 0–3 have already been written to Data Block 0 as shown in Fig. 49. The Log–Block scheme (see Section 5.3.1) without a write buffer updates LPNs 0, 0, 0, and 3 to Log Block 0, and then a full merge operation is performed as shown in Fig. 49A. Therefore, the Log-Block scheme without a write buffer requires 4 copy and 2 erase operations. On the other hand, the Log-Block scheme with a write buffer shown in Fig. 49B updates LPNs 0, 0, 0, and 3 to the write buffer in the SRAM instead of the flash memory. The write buffer holds the updated data until it is evicted and written back to the flash memory, which reduces write traffic and delay. Therefore, properly selecting a victim block is crucial for performance of SSDs and a number of schemes have been proposed [53–55].

The problem with only focusing on the victim block selection is that a nonsequential LPN order may cause similar number of copy and erase operations as in the case without a write buffer. For example, in Fig. 49B, only LPNs 0 and 3 need to be written to the Log Block 0 during eviction. However, a full merge operation will be required afterward due to nonsequential LPN order of the Log Block 0. For this reason, a number of methods have been proposed to reduce the number of full merge operations and/or erase operations.

The *Block Padding Least Recently Used* (BPLRU) scheme utilizes the *page padding* technique to reduce the number of full merge operations [52]. This technique pads the log block in the write buffer with valid pages from the data block in the flash memory as shown in Fig. 50. Suppose the same requests are issued as shown in Fig. 49 (i.e., updates to LPNs 0, 0, 0, and 3), then one of the blocks from the write buffer is evicted to Log Block 0 based on the LRU policy. The BPLRU scheme reads LPN1 and LPN2 to pad the evicted block, and then performs sequential writes to LPNs 0, 1, 2, and 3 in Log Block 0. Afterward, Log Block 0 becomes Data Block 0 by performing a simple switch operation instead of an expensive full merge operation. Therefore, the BPLRU scheme significantly improves the random write performance.

The *Effective Page Padding* technique enhances the endurance of SSDs by reducing the number of erase operations [46]. This is achieved by employing the *Evicted Block Table* to store the block number and valid page numbers of the evicted block. Figs. 51 and 52 provide a comparison between BPLRU and the Effective Page Padding technique. Suppose write to LPN 0, write to LPN 2, the first buffer eviction, write to LPN 3, and the second buffer eviction occur in order. The BPLRU scheme reads LPNs 1 and 3 to perform page padding and then performs sequential writes to LPNs 0–3 due to the first buffer eviction as shown in Fig. 51. The write request to LPN 3

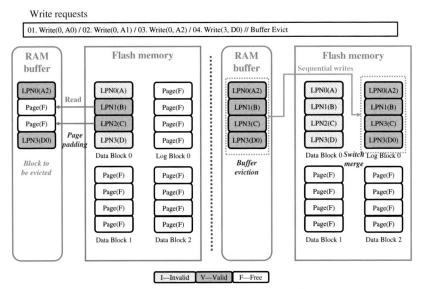

Fig. 50 The page padding technique.

(D0) will be held in the write buffer. Finally, BPLRU performs page padding when the second buffer eviction occurs. Therefore, BPLRU requires two erase operations.

On the other hand, the Effective Page Padding scheme performs partial page padding up to the highest cached LPN as shown in Fig. 52. For example, if LPNs 0 and 2 are cached in the write buffer, the highest cached LPN is 2. Thus, LPN 1 is read from the flash memory to perform partial page padding because it is lower than the highest cached LPN. Since LPNs higher than the highest cached LPN in Data Block 0 (i.e., LPN 3 in Data Block 0) are not involved in page padding, they are instead stored in the Evicted Block Table. Then, the sequential write requests to LPNs 0–2 are issued to Log Block 0 during the first buffer eviction as shown in Fig. 52. Afterward, the write to LPN 3 is stored in the write buffer. When the second buffer eviction occurs, LPN 3 is written to Log Block 0. Therefore, the Effective Page Padding scheme requires only one erase operation. However, if the highest cached LPN is close to the block size, full page padding can result in better performance than partial page padding. The Effective Page Padding scheme defines a threshold for a padding management. If the highest cached LPN is lower than the threshold, partial page padding will be performed. Otherwise, full page padding will be performed, which is the same as BPLRU. As a result, the Effective Page Padding scheme reduces the number of erase operations compared to BPLRU by up to 18.2%.

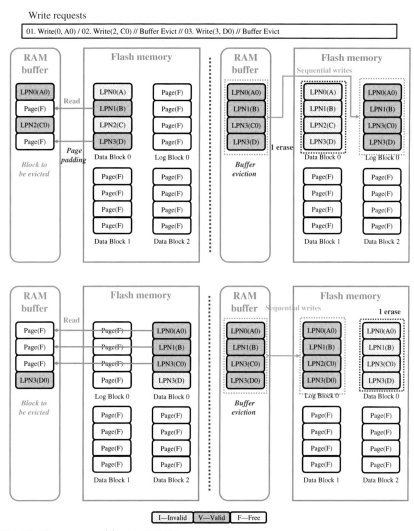

Fig. 51 The page padding in BPLRU.

There is also a method called *Predicted average Updated Distance Least Recently Used* (PUD-LRU), which distinguishes the entries in the write buffer as Frequently Updated Group (FUG) and Infrequently Updated Group (IUG) to exploit temporal locality [56]. FUG and IUG are classified based on the *update distance*, which is similar to the update interval used in LAST (see Section 5.3.4). Moreover, the PUD-LRU scheme chooses a victim block among the blocks in IUG in an LRU manner. Thus, PUD-LRU considers both frequency and proximity of accesses.

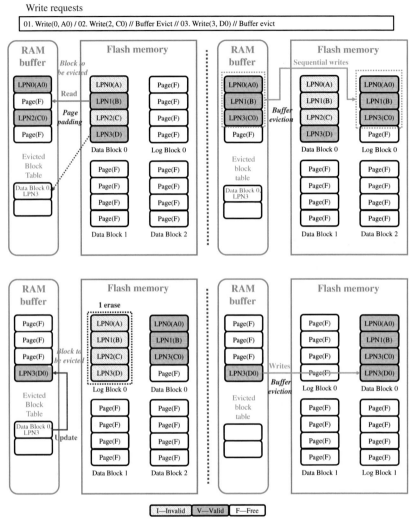

Fig. 52 The Effective Page Padding method.

11.1.8 Garbage collection aware striping

The *Round-robin* striping technique discussed in Section 9 is typically used to exploit channel-level concurrency, but it incurs extra garbage collection overhead. *Garbage collection aware striping* (GCAS) basically uses striping with *block scheduling* to improve garbage collection efficiency and thus performance [42]. Fig. 53 illustrates round-robin striping and GCAS. GCAS basically performs striping by grouping sequential logical pages in the same block.

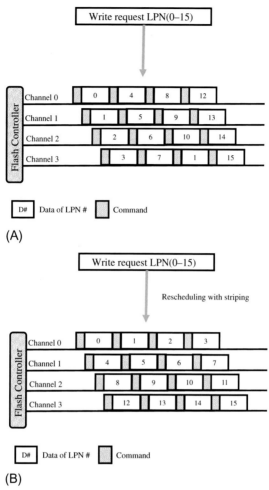

Fig. 53 Comparison for round-robin striping (A) and garbage collection aware striping (B).

Fig. 54 compares the garbage collection efficiency of the round–robin striping versus the GCAS. In this example, the SSD is assumed to have two channels each consisting of two blocks. When write requests occur to LPNs 0–7, the round-robin striping technique schedules writes to LPNs 0, 2, 4, and 6 to Block 0 and writes to LPNs 1, 3, 5, and 7 to Block 2 as shown in Fig. 54A. Afterward, another set of write requests to LPNs 0, 1, 2, and 3 causes LPNs 0 and 2 to be updated in Block 1 and LPNs 1 and 3 to be updated in Block 3. Since there are no free blocks, garbage collection is performed. Thus, LPNs 4 and 6 are copied to Block 1 and LPNs 5 and 7 are

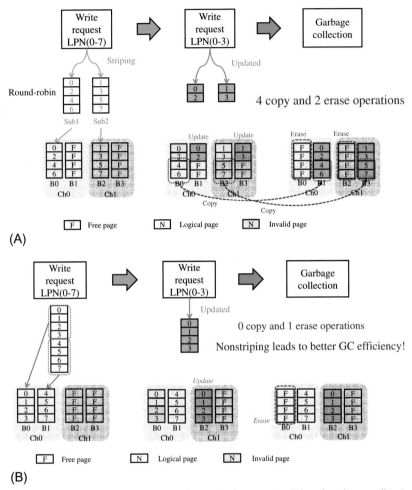

Fig. 54 Garbage collection efficiency of round-robin striping (A) and garbage collection aware striping (B).

copied to Block 3, and then Blocks 0 and 2 are erased. As a result, the round-robin striping technique requires four copy and two erase operations. In contrast, the block striping technique allocates LPNs 0, 1, 2, and 3 to Block 0 and LPNs 4, 5, 6, and 7 to Block 1 as shown in Fig. 54B. When another set of write requests to LPNs 0–3 occurs, only one erase operation is needed leading to better garbage collection efficiency.

11.1.9 Harey Tortoise
The *Harey Tortoise* technique improves not only the best-case performance for latency-critical operations but also their average performance [48]. This

is achieved with a flexible FTL that considers the fact that access time for the first bit of MLC is much faster than the access time for the second bit of MLC. Therefore, the basic idea behind the Harey Tortoise technique is to map the first and the second bit of MLCs to *fast pages* and *slow pages*, respectively. The authors in [48] show that fast pages are 4.8 times faster than slow pages. Therefore, the Harey Tortoise technique provides fast pages for latency-critical operations and slow pages to maximize memory density, which provides similar performance to SLC with higher density using MLC device.

11.1.10 Willow

Willow is a flexible, user-programmable SSD that allows for a customized implementation to improve the performance. This includes complex atomic operation, native caching operation, delegating storage allocation decisions to the SSD, and offloading file system permission checks to hardware [44]. According to [57], a customized SSD that offloads file system permissions reduces latency by up to 58% and improves throughput by 7.6 times.

Fig. 55A shows the architecture of a conventional SSD with multiple CPUs and nonvolatile memory (NVM), i.e., flash memory, connected to the host system via the nonvolatile memory express (NVMe) interface. Although a conventional SSD has processing units, they are not directly programmable from the host system.

Fig. 55B depicts the architecture of Willow SSD, which has *storage processor units* (SPUs) each consisting of a microprocessor and a network

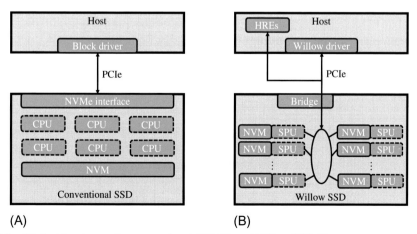

(A)   (B)

Fig. 55 Architectures of a conventional SSD (A) and Willow SSD (B).

interface to communicate with other SPUs. The *host remote procedure call end-points* (HREs) are a set of objects that allow the operating system and applications to communicate with SPUs. Therefore, SSD applications can be installed on the Willow driver, and then it creates and manages HREs. This flexibility allows for supporting a variety of functions to improve their performance. For example, Willow provides an OS-bypass interface in order to avoid overhead from system calls and file systems, which allows applications to perform READ() and WRITE() operations without OS intervention. The performance of SSDs can also be improved by an OS-bypass interface when running trusted codes in the SSDs. Willow can also be programmed to be used as a cache for a larger conventional storage system. Furthermore, SPUs in the Willow can handle data intensive computations, thereby reducing expensive data movement from SSD to the host. These features allow for a user programmable system that can extend the behavior of an SSD and therefore improve its performance [44].

11.2 Lifetime and reliability

Since lifetime and reliability are important issues, there have been many research efforts that focus on improving longevity and reliability of flash memories. Lifetime and reliability are highly related because the lifetime of flash memories is determined by the reliability of flash cells. If a flash memory is no longer reliable, the FTL considers its lifetime as being over. Typically, wear leveling techniques are employed in order to improve lifetime. On the other hand, SSDs rely on ECC and Bad Block Management in order to improve their reliability. Bose–Chaudhuri–Hocquenghem (BCH) codes and low-density parity-check (LDPC) codes are widely used as ECC for SSDs [1]. Skip Block and Reserve Block methods are generally employed for bad block management (see Section 8) [7]. Sections 11.2.1–11.2.6 discuss some recent techniques to improve lifetime and reliability.

11.2.1 Dynamic program and erase scaling

There is a tradeoff between the endurance of flash memories and the performance of P/E operations. Erasing a NAND block with a lower voltage causes electrons to penetrate the oxide layer more slowly and gently, and as a result, there is less damage to the oxide layer than the traditional high-voltage method, which significantly improves endurance. However, the performance degrades due to slower P/E operations. The *dynamic program and erase scaling* (DPES) scheme exploits this tradeoff by dynamically choosing between a high voltage for a short P/E operation or a low voltage for a long P/E operation [47]. DPES chooses the high voltage P/E operation when the number of outstanding write requests is high, and the low voltage P/E operation when the number of outstanding write requests is low.

11.2.2 Zombie NAND technique

The *Zombie NAND* technique exploits the gap in P/E cycles between TLC and MLC by resurrecting dead TLCs to MLCs. MLC and TLC have lifetimes of 10 k and 1 k P/E cycles, respectively. However, both MLC and TLC use exactly the same flash device structure except that TLC has more V_t levels than MLC. Since TLC becomes unreliable after 1 k P/E cycles, the V_t recognition process is changed to MLC allowing flash blocks to be used until 10 k P/E cycles. The Zombie NAND technique proposes a controlled *wear-unleveling* method as a way to resurrect dead blocks [17]. The wear-unleveling transforms TLC to MLC and MLC to SLC. Fig. 56 shows the advantage of the Zombie NAND technique. The capacity of TLC, MLC, and SLC decreases when the number of P/E cycles become above 1 k, 10 k, and 100 k, respectively. The Zombie NAND technique changes TLC to MLC after 1 k P/E cycles, and MLC to SLC after 10 k P/E cycles. Therefore, the resulting capacity follows the black line shown in Fig. 56, which extends the lifetime of flash memories. However, density is a tradeoff factor due to the reduction in the memory size.

11.2.3 Self-healing SSDs

As mentioned in Section 2.3, trapped charges inside the oxide layer increases V_t of a flash cell. The *self-healing SSDs* technique recovers these damaged

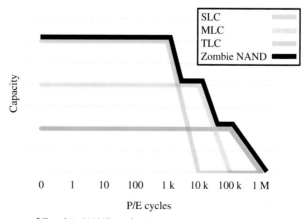

Fig. 56 Concept of Zombie NAND technique.

cells using a high temperature to repel the trapped charges from the oxide layer [45, 58]. This is illustrated in Fig. 57, where an extra heater die is used to create heat. Therefore, the lifetime is improved by solving the endurance problem from trapped charges [45]. However, the extra heater die requires additional die area and power.

11.2.4 Flash correct and refresh

As mentioned in Section 2.3, charged electrons inside the floating gate can escape due to SILC causing retention errors. The *flash correct-and-refresh* (FCR) technique introduces a new method to improve lifetime, which is illustrated in Fig. 58 [49]. As time passes, electrons escape from the floating gate causing V_t of the cell with data "0" becomes closer to the read reference voltage (V_{READ}). Therefore, the main idea of FCR is to periodically read and correct retention errors by performing in-place reprogramming. This is achieved by modeling the BER-based retention time for a damaged cell,

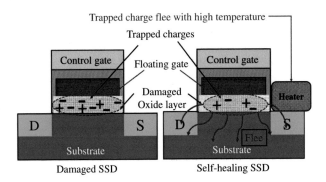

Fig. 57 Trapped charges and self-healing SSD.

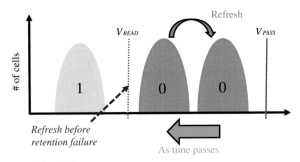

Fig. 58 Concept of the FCR.

and then periodically refreshing the current data state "0". This significantly improves the lifetime of flash memories and only requires modification to the SSD controller firmware. However, FCR requires constant power as well as extra processing to refresh flash cells. Furthermore, it cannot recover from retention failures that have already occurred during off power state.

11.2.5 Retention optimized reading and retention failure recovery

As discussed in Section 2.3, the state of data "1" cannot be properly read if the trapped charges causes V_t to shift and becomes very close to V_{READ}. This issue can be resolved using the *retention optimized reading* (ROR) technique [50]. Fig. 59 shows V_t as a function of the number of P/E cycles for MLC, which has three read reference voltages (V_{READ1}, V_{READ2}, and V_{READ3}). As can be seen, V_t of erased data "11" (blue line) and V_t of programmed data "01" (green line) increase as the number of P/E cycles increases. A read failure will occur when V_t reaches points A, and thus the flash memory is no longer considered reliable. In order to avoid read failures, the ROR technique changes the read reference voltage V_{READ3} to *Optimized V_{READ3}* before V_t reaches point A. Therefore, the lifetime of flash memories can be extended.

Even with ROR, the problem of charge leakage still exists. FCR discussed in Section 11.2.4 is one possible solution for SILC, but it cannot recover from retention errors that occurred during the power off state. The *retention failure recovery* (RFR) technique shown in Fig. 60 recovers data using a probabilistic method. This is done by classifying types of cells as *fast leaking* and *slow leaking*. In order to determine whether a cell is fast or slow

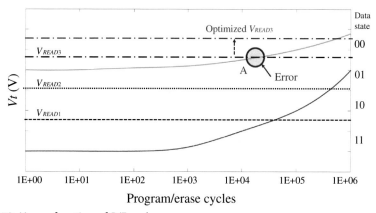

Fig. 59 V_t as a function of P/E cycles.

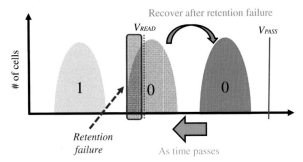

Fig. 60 Concept of the RFR.

leaking, all the flash cells are programmed with the same voltage level. Then, after a certain amount time, fast leaking cells and slow leaking cells can be distinguished based on whether the remaining level is lower or higher than the mean threshold voltage level, respectively. This way, the original state can be predicted to recover the data from retention failures.

11.2.6 BER-based wear leveling and bad block management

The reliability and lifetime of flash memories are typically measured by the number of P/E cycles. However, flash cells do not become damaged uniformly based only on the number of P/E cycles, thus bit error rate (BER) is a more practical factor for properly reflecting the lifetime of flash cells. The *BER-based wear leveling and bad block management* technique introduces a BER-based damage estimator to more precisely measure the damage level of flash cells [33]. The BER-based wear leveling improves reliability by relying on the same BER threshold for all the blocks. In other words, some blocks that have exceeded the P/E cycle limit but have a good BER can be used until it dips below a threshold. However, extra processing is required to update BER for each block incurring a performance overhead.

12. Conclusion

This survey presented a comprehensive coverage of issues in SSDs, including all the basic functionalities provided by FTL. Moreover, it provided the state of the art in SSD research on improving performance, lifetime, and reliability. Our survey also clearly shows that the research challenges in SSDs are vast and many of the issues presented are intricately intertwined. Therefore, dealing with these issues holistically will be crucial for improving the performance of future SSD systems.

References

[1] R. Micheloni, A. Marelli, K. Eshghi, Inside Solid State Drives (SSDs), vol. 37, Springer Science & Business Media, 2012.

[2] N. Agrawal, V. Prabhakaran, T. Wobber, J.D. Davis, M.S. Manasse, R. Panigrahy, Design tradeoffs for SSD performance, in: USENIX Annual Technical Conference, 2008, pp. 57–70.

[3] C. Dirik, B. Jacob, The performance of PC Solid State Disks (SSDs) as a function of bandwidth, concurrency, device architecture, and system organization, in: ACM SIGARCH Computer Architecture News, vol. 37, ACM, 2009, pp. 279–289.

[4] J.-U. Kang, H. Jo, J.-S. Kim, J. Lee, A superblock-based flash translation layer for NAND flash memory, in: Proceedings of the 6th ACM & IEEE International conference on Embedded software, ACM, 2006, pp. 161–170.

[5] Y.-H. Chang, J.-W. Hsieh, T.-W. Kuo, Endurance enhancement of flash-memory storage systems: an efficient static wear leveling design, in: Proceedings of the 44th Annual Design Automation Conference, ACM, 2007, pp. 212–217.

[6] E. Spanjer, Flash management—why and how? SMART Modular Technologies, 2009, p. 14.

[7] S.T. Microelectronics, Bad block management in NAND flash memories, 2004. Application note AN-1819, Geneva, Switzerland.

[8] T.-S. Chung, D.-J. Park, S. Park, D.-H. Lee, S.-W. Lee, H.-J. Song, System software for flash memory: a survey, in: International Conference on Embedded and Ubiquitous Computing, Springer, 2006, pp. 394–404.

[9] T.-S. Chung, D.-J. Park, S. Park, D.-H. Lee, S.-W. Lee, H.-J. Song, A survey of flash translation layer, J. Syst. Archit. 55 (5) (2009) 332–343.

[10] A.R. Olson, D.J. Langlois, Solid state drives data reliability and lifetime, 2008. Imation, White Paper.

[11] D. Kim, K. Bang, S.-H. Ha, S. Yoon, E.-Y. Chung, Architecture exploration of high-performance. PCs with a solid-state disk, IEEE Trans. Comput. 59 (7) (2010) 878–890.

[12] R. Bez, E. Camerlenghi, A. Modelli, A. Visconti, Introduction to flash memory, Proc. IEEE 91 (4) (2003) 489–502.

[13] R. Micheloni, L. Crippa, A. Marelli, Inside NAND Flash Memories, Springer Science & Business Media, 2010.

[14] A. Hiraiwa, T. Saito, A. Daicho, H. Kawarada, Accuracy assessment of sheet-charge approximation for Fowler-Nordheim tunneling into charged insulators, J. Appl. Phys. 114 (13) (2013) 134501.

[15] M. Momodomi, T. Tanaka, Y. Iwata, Y. Tanaka, H. Oodaira, Y. Itoh, R. Shirota, K. Ohuchi, F. Masuoka, A 4 Mb NAND EEPROM with tight programmed V_t distribution, IEEE J. Solid-State Circuits 26 (4) (1991) 492–496.

[16] Micron, Micron NAND Flash Memory datasheet MT29F4G08AAA, MT29F4G08BAA, MT29F4G08DAA, MT29F16G08FAA, 2007. Micron Technology, Inc.

[17] E.H. Wilson, M. Jung, M.T. Kandemir, ZombieNAND: resurrecting dead NAND flash for improved SSD longevity, in: 2014 IEEE 22nd International Symposium on Modelling, Analysis & Simulation of Computer and Telecommunication Systems, IEEE, 2014, pp. 229–238.

[18] N. Mielke, H.P. Belgal, A. Fazio, Q. Meng, N. Righos, Recovery effects in the distributed cycling of flash memories, in: 2006 IEEE International Reliability Physics Symposium Proceedings, IEEE, 2006, pp. 29–35.

[19] Y.-B. Park, D.K. Schroder, Degradation of thin tunnel gate oxide under constant Fowler-Nordheim current stress for a flash EEPROM, IEEE Trans. Electron Devices 45 (6) (1998) 1361–1368.

[20] A. Modelli, A. Visconti, R. Bez, Advanced flash memory reliability, in: International Conference on Integrated Circuit Design and Technology, 2004, ICICDT'04, IEEE, 2004, pp. 211–218.

[21] M. Jung, W. Choi, S. Srikantaiah, J. Yoo, M.T. Kandemir, HIOS: a host interface I/O scheduler for solid state disks, ACM SIGARCH Comput. Archit. News 42 (3) (2014) 289–300.
[22] J. Kim, J.M. Kim, S.H. Noh, S.L. Min, Y. Cho, A space-efficient flash translation layer for compact flash systems, IEEE Trans. Consum. Electron. 48 (2) (2002) 366–375.
[23] X.-Y. Hu, E. Eleftheriou, R. Haas, I. Iliadis, R. Pletka, Write amplification analysis in flash-based solid state drives, in: Proceedings of SYSTOR 2009: The Israeli Experimental Systems Conference, ACM, 2009, p. 10.
[24] S.-W. Lee, D.-J. Park, T.-S. Chung, D.-H. Lee, S. Park, H.-J. Song, A log buffer-based flash translation layer using fully-associative sector translation, ACM Trans. Embed. Comput. Syst. (TECS) 6 (3) (2007) 18.
[25] S. Lee, D. Shin, Y.-J. Kim, J. Kim, LAST: locality-aware sector translation for NAND flash memory-based storage systems, ACM SIGOPS Oper. Syst. Rev. 42 (6) (2008) 36–42.
[26] A. Gupta, Y. Kim, B. Urgaonkar, DFTL: A Flash Translation Layer Employing Demand-Based Selective Caching of Page-Level Address Mappings, vol. 44, ACM, 2009.
[27] J. Hu, H. Jiang, L. Tian, L. Xu, GC-ARM: garbage collection-aware RAM management for flash based solid state drives, in: 2012 IEEE 7th International Conference on Networking, Architecture and Storage (NAS), IEEE, 2012, pp. 134–143.
[28] A.I. Alsalibi, P. Sumari, S.A. Alomari, M.A. Al-Betar, Performance and reliability concern scheme for efficient garbage collection and wear leveling on flash memory-based solid state disk, Microsyst. Technol. (2016) 1–15.
[29] B. Van Houdt, Performance of garbage collection algorithms for flash-based solid state drives with hot/cold data, Perform. Eval. 70 (10) (2013) 692–703.
[30] W. Bux, I. Iliadis, Performance of greedy garbage collection in flash-based solid-state drives, Perform. Eval. 67 (11) (2010) 1172–1186.
[31] L.-P. Chang, On efficient wear leveling for large-scale flash-memory storage systems, in: Proceedings of the 2007 ACM symposium on Applied computing, ACM, 2007, pp. 1126–1130.
[32] M. Murugan, D.H.C. Du, Rejuvenator: a static wear leveling algorithm for NAND flash memory with minimized overhead, in: 2011 IEEE 27th Symposium on Mass Storage Systems and Technologies (MSST), IEEE, 2011, pp. 1–12.
[33] B. Peleato, H. Tabrizi, R. Agarwal, J. Ferreira, BER-based wear leveling and bad block management for NAND flash, in: 2015 IEEE International Conference on Communications (ICC), IEEE, 2015, pp. 295–300.
[34] M. Jung, E.H. Wilson, D. Donofrio, J. Shalf, M.T. Kandemir, NANDFlashSim: intrinsic latency variation aware NAND flash memory system modeling and simulation at microarchitecture level, in: 012 IEEE 28th Symposium on Mass Storage Systems and Technologies (MSST), IEEE, 2012, pp. 1–12.
[35] M. Jung, M.T. Kandemir, An evaluation of different page allocation strategies on high-speed SSDs, in: HotStorage, 2012.
[36] Y. Hu, H. Jiang, D. Feng, L. Tian, H. Luo, S. Zhang, Performance impact and interplay of SSD parallelism through advanced commands, allocation strategy and data granularity, in: Proceedings of the International Conference On Supercomputing, ACM, 2011, pp. 96–107.
[37] Y. Wang, Z. Qin, R. Chen, Z. Shao, Q. Wang, S. Li, L.T. Yang, A Real-Time Flash Translation Layer for NAND Flash Memory Storage Systems, IEEE Trans. Multi-Scale Comput. Syst. 2 (1) (2016) 17–29.

[38] J. Kim, D.H. Kang, B. Ha, H. Cho, Y.I. Eom, MAST: multi-level associated sector translation for NAND flash memory-based storage system, in: Computer Science and Its Applications, Springer, 2015, pp. 817–822.
[39] Q. Wei, C. Chen, M. Xue, J. Yang, Z-MAP: a zone-based flash translation layer with workload classification for solid-state drive, ACM Trans. Storage (TOS) 11 (1) (2015) 4.
[40] P. Dongchul, B. Debnath, H.C.D. David, A dynamic switching flash translation layer based on page-level mapping, IEICE Trans. Inform. Syst. 99 (6) (2016) 1502–1511.
[41] L.-P. Chang, Y.-S. Liu, W.-H. Lin, Stable greedy: adaptive garbage collection for durable page-mapping multichannel SSDs, ACM Trans. Embed. Comput. Syst. (TECS) 15 (1) (2016) 13.
[42] M. Huang, Y. Wang, Z. Liu, L. Qiao, Z. Shao, A garbage collection aware stripping method for solid-state drives, in: The 20th Asia and South Pacific Design Automation Conference, IEEE, 2015, pp. 334–339.
[43] Q. Zhang, X. Li, L. Wang, T. Zhang, Y. Wang, Z. Shao, Lazy-RTGC: a real-time lazy garbage collection mechanism with jointly optimizing average and worst performance for NAND flash memory storage systems, ACM Trans. Des. Autom. Electron. Syst. (TODAES) 20 (3) (2015) 43.
[44] S. Seshadri, M. Gahagan, M.S. Bhaskaran, T. Bunker, A. De, Y. Jin, Y. Liu, S. Swanson, Willow: a user-programmable SSD, in: OSDI, 2014, pp. 67–80.
[45] Q. Wu, G. Dong, T. Zhang, A first study on self-healing solid-state drives, in: 2011—3rd IEEE International Memory Workshop (IMW), IEEE, 2011, pp. 1–4.
[46] E. Ogawa, K. Kise, An Effective Page Padding Method for RAM Buffer Algorithms to Enhance the SSD Endurance, in: 2016 Fourth International Symposium on Computing and Networking (CANDAR), IEEE, 2016, pp. 133–139.
[47] J. Jeong, S.S. Hahn, S. Lee, J. Kim, Lifetime improvement of NAND flash-based storage systems using dynamic program and erase scaling, in: Proceedings of the 12th USENIX Conference on File and Storage Technologies (FAST 14), 2014, pp. 61–74.
[48] L.M. Grupp, J.D. Davis, S. Swanson, The harey tortoise: managing heterogeneous write performance in SSDs, in: USENIX Annual Technical Conference, 2013, pp. 79–90.
[49] Y. Cai, G. Yalcin, O. Mutlu, E.F. Haratsch, A. Cristal, O.S. Unsal, K. Mai, Flash correct-and-refresh: retention-aware error management for increased flash memory lifetime, in: 2012 IEEE 30th International Conference on Computer Design (ICCD), IEEE, 2012, pp. 94–101.
[50] Y. Cai, Y. Luo, E.F. Haratsch, K. Mai, O. Mutlu, Data retention in MLC NAND flash memory: characterization, optimization, and recovery, in: 2015 IEEE 21st International Symposium on High Performance Computer Architecture (HPCA), IEEE, 2015, pp. 551–563.
[51] J. Kim, Y. Oh, E. Kim, J. Choi, D. Lee, S.H. Noh, Disk schedulers for solid state drivers, in: Proceedings of the Seventh ACM international Conference on Embedded Software, ACM, 2009, pp. 295–304.
[52] H. Kim, S. Ahn, BPLRU: a buffer management scheme for improving random writes in flash storage, in: FAST, vol. 8, 2008, pp. 1–14.
[53] S.-Y. Park, D. Jung, J.-U. Kang, J.-S. Kim, J. Lee, CFLRU: a replacement algorithm for flash memory, in: Proceedings of the 2006 International Conference on Compilers, Architecture and Synthesis for Embedded Systems, ACM, 2006, pp. 234–241.
[54] H. Jo, J.-U. Kang, S.-Y. Park, J.-S. Kim, J. Lee, FAB: flash-aware buffer management policy for portable media players, IEEE Trans. Consum. Electron. 52 (2) (2006) 485–493.

[55] S. Jiang, X. Ding, F. Chen, E. Tan, X. Zhang, DULO: an effective buffer cache management scheme to exploit both temporal and spatial locality, in: Proceedings of the 4th Conference on USENIX Conference on File and Storage Technologies, vol. 4, 2005, p. 8.
[56] J. Hu, H. Jiang, L. Tian, L. Xu, PUD-LRU: an erase-efficient write buffer management algorithm for flash memory SSD, in: 2010 IEEE International Symposium on Modeling, Analysis & Simulation of Computer and Telecommunication Systems (MASCOTS), IEEE, 2010, pp. 69–78.
[57] A.M. Caulfield, T.I. Mollov, L.A. Eisner, A. De, J. Coburn, S. Swanson, Providing safe, user space access to fast, solid state disks, ACM SIGARCH Comput. Archit. News 40 (1) (2012) 387–400.
[58] V. Mohan, T. Siddiqua, S. Gurumurthi, M.R. Stan, How I learned to stop worrying and love flash endurance, HotStorage 10 (2010) 3.

About the authors

Youngbin Jin received his B.S. degree in Electrical Engineering from Washington State University in 2010 and his M.S. degree in Electrical Engineering from the University of Texas at Dallas in 2013. He is currently pursuing a Ph.D. degree in Electrical and Computer Engineering at Oregon State University. His research involves cross-level issues between hardware and software and his interests include Computer Architecture, Computer Network, Multimedia, VLSI design, FPGAs, and Memory Systems.

Prof. Ben Lee received his B.E. degree in Electrical Engineering in 1984 from the Department of Electrical Engineering at State University of New York (SUNY) at Stony Brook, and his Ph.D. degree in Computer Engineering in 1991 from the Department of Electrical and Computer Engineering at the Pennsylvania State University. He received the Loyd Carter Award for Outstanding and Inspirational Teaching in 1994, the Alumni Professor Award for Outstanding Contribution to the College and the University from the OSU College of Engineering in 2005, and the HKN Innovation Teaching Award from Eta Kappa Nu, School of Electrical Engineering and Computer Science in 2008. He has been on the program committees and organizing committee for numerous international conferences including IEEE Consumer Communications & Networking Conference (CCNC), and IEEE International Conference on Pervasive Computing and Communications (PerCom), and IEEE Workshop on Pervasive Wireless Networking (PWN). He was the TPC-Chair for the 15th Annual IEEE Consumer Communications & Networking Conference (CCNC 2018). He was a Guest Editor for the Special Issue on "Wireless Networks and Pervasive Computing" for the Journal of Pervasive Computing and Communications (JPCC). He was also an invited speaker at the 2007 International Conference on Embedded Software and System and a Keynote Speaker at the 2014 ACM International Conference on Ubiquitous Information Management and Communication. He is currently the General Chair for the 17th Annual IEEE Consumer Communications & Networking Conference (CCNC 2020). He is also an Adjunct Faculty member at Korea Advanced Institute of Science and Technology (KAIST). His research interests include multimedia streaming, wireless networks, embedded systems, computer architecture, multithreading and thread-level speculation, and parallel and distributed systems.

CHAPTER TWO

Revisiting VM performance and optimization challenges for big data

Muhammad Ziad Nayyer[a,b], Imran Raza[b], Syed Asad Hussain[b]
[a]Department of Computer Science, GIFT University, Gujranwala, Pakistan
[b]Department of Computer Science, Communication and Network Research Centre, COMSATS University Islamabad, Lahore, Pakistan

Contents

Abstract

The concept of virtualization in cloud computing aims to maximize resource utilization and minimize cost by deploying multiple Virtual Machines (VMs) on a single physical server sharing resources such as CPU, Cache, I/O, and Memory. The sharing of these resources can cause severe performance degradation, thus requiring VM migration techniques for performance enhancement. The introduction of big data has made the performance enhancement more challenging due to extensive volume, velocity,

Advances in Computers, Volume 114
ISSN 0065-2458
https://doi.org/10.1016/bs.adcom.2019.02.002

variety, variability, and veracity of data. The existing VM performance and optimization challenges need to be optimized for big data use cases. This chapter presents big data triggered VM performance challenges focusing big data applications and storage migration in cloud computing. State of the art VM migration techniques are evaluated against challenges posed by big data to outline possible solutions and research challenges.

1. Introduction

Big data refer to the volume of data that are challenging to handle due to its volume, velocity, variety, variability, and veracity. Frequency of the data generation has increased massively in the last decade demanding more computing resources to handle operations such as data capture, data storage, data analysis, and data visualization [1]. More computing resources mean more power utilization and complexity in terms of distributed algorithms and applications necessary to handle big data. Provisioning and handling such resources at the local level is costly and difficult. Therefore, organizations are seeking third party services. Cloud computing provides a viable solution for rendering infrastructure and resources by offering different service oriented model such as Infrastructure as a Service (IaaS), Platform as a Service (PaaS) and Software as a Service (SaaS). Cloud computing make use of virtualization technology to maximize resource utilization with minimum cost. International Data Corporation (IDC) indicates that cloud computing will generate a revenue up to $53.1 billion by 2019 [2]. Cloud Service Providers (CSPs) use data centers to house cloud services and cloud-based resources. For cloud-hosting purposes, vendors often use multiple data centers in several geographic locations to ensure data availability during outages and other data center failures. However, there are various challenges faced by CSPs which include energy efficiency, seamless services, provisioning of the resource rich servers, cooling, high speed storage, and virtualization related challenges. Virtualization related challenges include rapid deployment, snapshotting, scheduling, prioritizing which may lead to resource contention. Provisioning more resources is a solution but not cost effective. VM migration on the other hand, is a more viable solution that is widely used. VM migration means that a VM facing performance issues due to limitation of resources (e.g., CPU, memory, storage, and network) is migrated from a single physical server to another resource rich server instead of provisioning more resource on the same server. This scaling technique requires less labor, time, and cost. In addition, it gives an advantage of resource provisioning on runtime with minimum downtime. There are various methods of VM migration that are discussed in Section 3.2.

A VM usually comprise of disk image and memory state. However, in data centers, a central storage called Network Attached Storage (NAS) is used for storing data. As per volume of the data, a number of NAS devices may be used. Big data applications deployed on VMs are connected with the NAS devices over a high speed LAN. When a certain migration is triggered over the Wide Area Network (WAN) and a VM is migrated from one data center to another, it's affiliated storage must also be migrated with the VM. Migration of affiliated storage is essential for successful working of the migrated VM in the new environment and is a major challenge as the volume of the data is very large. Efficient algorithms are required to reduce data size, network traffic, and demand for bandwidth while preserving energy [3,4]. Additionally, a fast, reliable, and secure network is also required for the migration of NAS. Hence, big data networks are the only solution to survive the challenges of future Internet [5]. The term "Big Data Network" is relatively new and being used rarely in the literature. However, it refers to the readiness of networks for big data [6–13]. Many big data network architectures have been proposed based on Software Defined Networks (SDNs) and Hadoop by different vendors such as Juniper [14], IBM [15], CISCO [16], and Oracle [17]. SDNs employ the abstraction of lower level functionality that separates the control and data plane to achieve both the programmability and speed of data transfer [18]. Hadoop is a platform providing structure for big data that offers data and content management along with warehousing functionalities [19], but state of the art virtualization techniques is yet to be equipped with Hadoop and SDN solutions.

It is imperative to analyze the existing techniques used for VM performance enhancement from big data perspective. We have classified hardware parameters and operational issues effecting the VM performance. This will help us to focus on the major points for the improvement of existing techniques as well as for the development of new ones. The contribution of this study can be segregated in two parts: In the first part, we have assessed performance enhancement techniques according to the classified performance parameters and operational challenges. In the second part, we have analyzed the same performance enhancement techniques from big data perspective and also proposed a future cloud model ready to handle big data challenges.

Rest of the chapter is organized as follows: Conventional VM performance issues in a single server, single data center and geographically distributed data center from big data perspective have been discussed in Section 2. Existing overhead mitigation techniques and their tradeoffs have been presented for big data in Section 3, future cloud model, insights and open research challenges have been mentioned in Section 4, and finally Section 5 concludes the chapter.

2. Performance issues from big data perspective

The world has moved from single server virtualization to single large data center and to geo-distributed data center. We have categorized the virtualization challenges from these three perspectives as shown in Fig. 1.

However, there are some common operational challenges as shown in Table 1.

Detail of performance challenges is as follows:

1) VM performance can be limited by poor prioritizing. Issues arise when the element of priority is not considered while performing resource allocation, thus increasing the risk of performance overhead [20].

2) Scanning for viruses and malware on host or guest OS also causes performance degradation [21,22] as it is a CPU intensive task. The duration of resource occupation is highly dependent upon the size of host and guest OS along with the data.

3) Some operating systems (OS) do offer power saving option which reduces the processor clock speed causing performance issues [21].

4) Migration operation brings substantial overhead on collocated VMs at source and destination servers as it utilizes computing and network resources. A poorly scheduled migration operation can severely degrade the performance. Concurrent migrations may also bring significant performance overhead [3].

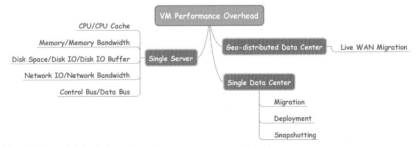

Fig. 1 Consolidated view of VM performance overheads.

Table 1 Performance overheads considering different operations.

Formation	Operations						
	Prioritizing	Antivirus/anti-malware scanning	Power saving	Scheduling	VM migration	Storage migration	Rapid deployment	Rapid snapshotting
Single server	✓	✓	✓	✓	–	–	✓	✓
Single data centre	✓	✓	✓	✓	✓	–	✓	✓
Multiple geo-distributed data centre	✓	✓	✓	✓	✓	✓	✓	✓

5) Timing and scheduling are two other factors that must be considered for operations such as VM snapshotting and migration. A poorly scheduled snapshotting or migration task may not only affect the VM but also severely affects the collocated VMs [21].

6) The host operating system's performance also affects the VM. Disk speed, cache, and bus are included in those factors effected by host OS. In Redundant Array of Independent Drive (RAID), if the host OS and VMs are on the same spindle then disk cache and bus sharing can cause significant overhead to the VM [23]. If a server gets low on memory, the operating system makes heavy use of virtual memory and the constant paging operation cause tremendous load on disk I/O resulting in performance degradation [24].

2.1 Single server issues

The basic goal of virtualization is to multiplex physical server's resources such as CPU, memory, and storage to provide better resource utilization as shown in Fig. 2.

The efforts for single server virtualization are focused on the isolation of shared resources such as CPU cache, control and data bus, memory bandwidth, disk I/O, and disk I/O buffer which are the major cause of contention, resulting in a performance overhead [25]. There are various hypervisors and VM monitors such as Citrix XenServer, VMware vSphere, and Microsoft Hyper-V [26]. The hypervisors are proven to be good in providing

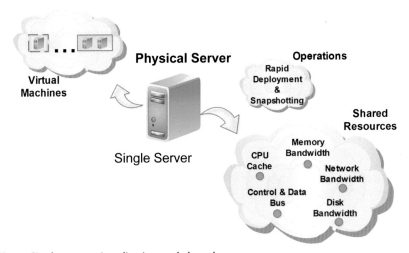

Fig. 2 Single server virtualization and shared resources.

resource isolation for multiple VMs. Isolation is achieved through full fledge CPU schedulers (e.g., Credit Scheduler in Xen [27]) and use static disk partitions. Routine operations such as rapid deployment and snapshotting which are very vital for the smooth execution of a VM are also a source of performance overhead. However, the effects of deployment and snapshotting are restricted within the boundary of a single server. The impact of all aforementioned factors is also linked with the nature of the workload. Workloads can be classified on the basis of resource intensiveness in terms of CPU, I/O, and memory, etc. In the presence of big data, the problems induced by the shared resources are increased drastically. Timely completion of larger jobs having increased computations and read/write operations require more supporting hardware in the form of CPU, memory, I/O, bus speed, bandwidth, and cache, etc.

2.2 Single data center issues

In the single data center, VMs are migrated over Local Area Network (LAN) from one server to another for performance improvement as shown in Fig. 3. The migration operation effects the performance of collocated VMs on both source and destination servers due to resource contention among migrated and already deployed VMs.

When a large volume of data is to be processed for complex jobs, it is highly desired by customers to lease a group of VMs from cloud service providers. MapReduce [28] is an example of such a complex job that requires extraordinary resources to run distributed algorithms using multiple VMs within a single data center. A single VM migration from the leased group of VMs can affect the overall performance due to the dependency among group VMs. Other examples may include scientific analysis and simulations that require more than a single VM's resources. From aforementioned

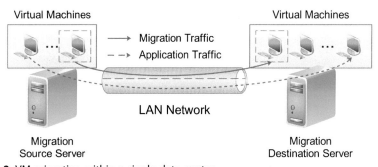

Fig. 3 VM migration within a single data center.

examples, the question arises that whether the techniques used to reduce performance overhead for a single server can also work for a single data center? The answer to this question is "No" as involvement of LAN and requirement for VM migration for load balancing does not persist for a single server. Big data demands for more resources in terms of storage, network processing, and bandwidth. A big data flow can affect the performance of a data center as it requires more network bandwidth and may trigger additional migrations for the handling and management of big data. Hence, the impact of big data challenges is more obvious in a data center environment than for a single server. We have discussed single data center issues such as shared resources, migration effects on collocated VMs, rapid deployment, and snapshotting in this section as shown in Fig. 4.

2.2.1 Migration

VM migration technique is commonly used for saving power and managing load in a cloud data center, but incurs a huge performance overhead which is unavoidable especially in case of concurrent live migrations [29]. The volume of data and supporting infrastructure is also an issue if speedy migration is required. The effects of concurrent deployments can also be seen on source and destination servers, including the applications running on the migrating VM itself. Migration and deployment operations further limits the CPU and network resources resulting in a longer migration time [30]. In LAN, where storage is not migrated as shared storage devices are used in the form

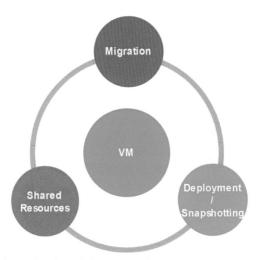

Fig. 4 Single and geo-distributed data center issues.

of independent NAS device, storage is not considered a challenging factor from big data point of view. However, active connections are closed and re-established from new destination. If a data intensive application having greater number of read and write operations is running on a VM that is in the middle of a migration process, then termination and re-establishment of active connections will result in a service disruption. There are two commonly used migration types, cold migration and live migration. In cold migration, a powered off VM and in live migration, a running VM is transferred from the source to destination server. Both techniques have their own pros and cons. In cold migration, the services running on the server are suspended during migration. The VM is first powered off and then migrated. The VM needs to be powered on after migration on the destination server which also takes some time lingering on the services running on the VM. In live migration, the static part (not changing or having any active write operation) of the VM is transferred first while VM keeps running on the source server. Then VM is suspended and the remaining part is transferred. Live VM migration reduces the down time but engages more resources for a longer time. Existing solutions of VM migration have different tradeoffs such as interference of live migration on collocated VMs against more CPU utilization [3], uninterrupted services during migration against more network and CPU load [3], and performance against more energy requirement [4]. There are many variants of both techniques which are further discussed in Section 3. All approaches try to optimize performance and achieve maximum utilization.

2.2.2 Deployment and snapshotting

Concurrent deployment and snapshotting are two consistently performed operations in a data center [31]. However, the nature of deployment and snapshotting operations in a data center is different from a single server. For the execution of complex and lengthy jobs having a large volume of data to be transmitted, it is very vital for such jobs to be timely executed using the fast deployment method. The fast deployment method requires a number of VMs to be initialized concurrently in cloud environment using more than one physical server [32]. The initialized VMs altogether work as a cluster and their combined resources are used to execute complex jobs. The cluster of VMs requires a tight coupling of resources through LAN and may use shared storage. VM snapshots are also maintained on the shared storage thus increasing the network dependency. Performance of a single VM on a physical server belonging to a cluster effects the other VMs on different physical servers belonging to the same cluster due to dependency of computations.

The main reason for the performance overhead is the large volume of data that are generated during rapid deployment and snapshotting operations. This big data result in a massive amount of network traffic interfering with regular data center traffic.

2.2.3 Shared resources

In the cloud environment, the services are provided in a shared model. Resources such as CPU cache, disk I/O and buffer, network I/O and memory bandwidth are shared among VMs. However, nature of these shared resources in a data center is changed as compared to a single server. The most critical shared resource in a data center is network bandwidth and I/O. The challenge of network bandwidth and I/O contention arises mostly at network storage level due to multiple VMs sending requests simultaneously. The I/O contention is dependent upon the volume of data coming in and going out of the storage device. In case of big data, where data are stored on multiple shared devices, the I/O contention will be at every device level resulting in a delayed operation. The latest statistics by Amazon EC2 confirm that the throughput variation faced by Standard Medium Instances (SMIs) can be up to 65% of the total network I/O [33], and the write I/O bandwidth can vary by as much as 50% from the mean [34].

2.3 Multiple geo-distributed data center issues

All issues discussed for the single data center are also valid for geo-distributed data centers. However, migration of VMs in a geo-distributed data center is treated differently due to the involvement of WAN. In the WAN environment, not only VM's disk and memory state is transferred but affiliated storage is also transferred [35]. The task of transferring affiliated storage becomes difficult due to the challenges of handling large data size, network traffic, link condition, available bandwidth, downtime of the VM in migration process, and energy consumption. An example depicting VM migration in WAN environment has been produced in Fig. 5.

The aforementioned challenges not only affect the source and destination data centers, but also the migrating VM itself and services running on it. Solutions for the single server or single data center do not work for geo-distributed data center environment due to the increased network volatility. The study in Ref. [36] has mentioned an experimental migration performed over WAN between the data centers in Illinois and Texas. The link was configured with a maximum throughput of 85 Mbps and VM was configured with 1.7 GB memory and 10 GB disk size. A web application

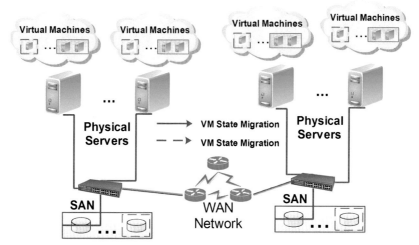

Fig. 5 VM across multiple geo-distributed data centers over WAN.

was hosted on the VM. The experimental results showed that it took around 40 min for disk transfer and 210 ms for memory transfer. During the migration process, the response time of web services was degraded to 52 ms on average, whereas, it was recorded as 10 ms before migration. The migration time increases with the increase in disk size. However, no storage was migrated in the experimental migration. In the presence of big data where a VM itself may consist of several disks due to distributed applications and algorithms running on it and have a number of affiliated storage drives, the impact on time will be drastic.

There is no doubt that the live migration technique used for VMs to move from one data center to another is very useful in reducing the services downtime, but it suffers from latency and bandwidth issues. However, big enterprises and IT firms are getting benefits from VM migration technique to transfer a single or number of VMs from a heavy-loaded data center to a light-loaded data center. The purpose of VM migration is to balance the load, save energy and enhance the resource utilization of the light loaded data center. The same technique is also used to move a single or number of VMs to a more powerful data center having high computing resources when the resource need of a VM is not being fulfilled. The cloud "Bursting" technique gives opportunity for small scale business companies and enterprises to jump between private and public clouds whenever their resource need is increased and are charged only for that period [36].

Table 2 summarizes all the major factors effecting the performance of single server, single data center and geo-distributed center.

Table 2 Virtual Machine performance overhead factors for different configurations.

Configurations	Shared resources	Deployment	Snapshot	VM migration	Network latency	Storage migration
				Operational problems		
Single server	✓	✓	✓	–	–	–
Single data centre	✓	✓	✓	✓	–	–
Multiple geo-distributed data centre	✓	✓	✓	✓	✓	✓

2.4 VM performance overhead evaluation

Three types of metrics are used for the evaluation of VM performance overhead. First is performance degradation metric, which can be calculated by comparing the performance degradation of various applications running in an IaaS cloud with the application running in an isolated environment.

$$\text{Performance Degradation} = \frac{\text{VM(IaaS)} - \text{VM(Isolation)}}{\text{VM(Isolation)}} \qquad (1)$$

Where, VM(IaaS) is VM in an IaaS cloud and VM(Isolation) is VM in an isolated environment. The larger magnitude of performance degradation indicates severity of the VM performance overhead. The second metric represents the variation in VM performance over an interval of Δt. VM performance variation can be represented by the coefficient of variation [37], and is formulated as

$$P_{v\Delta t} = \frac{1}{\overline{x}} \sqrt{\frac{1}{n_x - 1} \sum_{i=1}^{n_x} (x_i - \overline{x})^2} \qquad (2)$$

where, x_i represents performance measured for VM, $P_{v\Delta t}$ represents variation in VM performance, n_x represents the number of performance measurements taken and $\overline{x}$ represents average VM performance over time Δt. The magnitude of performance variation directly corresponds to server's performance overhead. It is very difficult to get the actual performance degradation measures due to security and measurement cost involved. Probing hardware and software requires additional tools and permission to extract performance parameters. When both discussed metrics fail to provide an accurate measure due to unavailability of tools to communicate with the hardware and software at the lower level, a third type of metric is used to measure the performance overhead, which considers the performance measurement of routine operations during live migration such as the amount of data to be transmitted, network traffic, link and bandwidth condition, downtime of the VM being recently migrated, and energy consumption.

3. Migration overhead mitigation techniques from big data perspective

Numerous approaches to minimize the VM performance overhead and optimize the recourses in a cloud environment have been proposed. These overhead mitigation and optimization approaches have been categorized according to single-server, single data center, and geo-distributed data center, respectively, and their overview have been provided in this chapter from big data perspective.

3.1 Single server techniques

Recent hypervisors such as Microsoft Hyper-V and XenServer [34] provide isolation for CPU, memory, and disk resources, but lack the capability of isolating the resources such as CPU cache, network I/O, memory bandwidth, and disk I/O. Many techniques have been proposed to overcome the resource contention problem and minimize the VM performance overhead on a single server. We further divide these techniques in two categories: First is based on collocated VM resource isolation and second is based on VM assignment optimization.

3.1.1 Collocated VM resource isolation

Resource isolation is a very common technique to avoid resource contention among collocated VMs on a single server. However, the resources that are hard to isolate such as CPU cache and memory bandwidth can also be handled to some extent using techniques such as Q-Cloud [38]. For better understanding, we have categorized the resource isolation approaches according to targeted resources of CPU, memory, cache, and I/O.

3.1.1.1 CPU targeted approaches

Q-Cloud uses an online feedback mechanism to build a model using Multi Input Multi Output (MIMO) to capture the performance interference interactions. The captured interference information is further used to perform closed loop resource management, which dynamically identifies

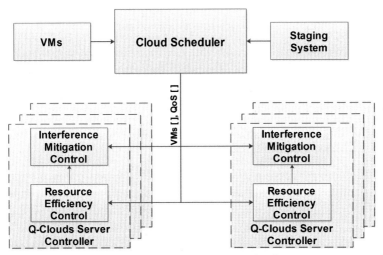

Fig. 6 VM deployment and QoS management with Q-Clouds.

the underutilized CPU resources. Application's resource need is provided by workload SLA and then required resources are dynamically provisioned. An application specific metric is defined in the SLA and semantic-less feedback signals are used for the online adaptation. A structural view of Q-Cloud is shown in Fig. 6. When we analyze aforementioned approach from the big data point of view, we can see that the online feedback consumes much bandwidth and suffers from latency issues considering the large volume of data. The frequency of performance interference interactions will also be increased resulting in delayed provisioning of resources making Q-Cloud technique unsuitable for big data applications.

3.1.1.2 Cache targeted approaches
Partitioning the cache space is a solution to overcome the contention issues among cache resources. It uses a coloring scheme to assign each virtual and physical memory page a different color. Only a same colored virtual page can be assigned to a physical page [39]. In the presence of big data, the process of partitioning cache space will become difficult as cache space is very limited and with the increase of data variation, the frequency of cache change will also be increased. Hence, this will result in increased frequency of mapping between virtual pages and physical memory pages consuming more CPU cycles causing an overhead.

A copy-based cache service is being introduced in Ref. [40] that improves the read performance by conserving a working set in shared memory with low management cost. There are two conditions for the cooperative caching scheme that system must use shared storage for all clients and cooperative cache hit rate must be greater than disk access. The cooperative cache scheme has been implemented using Xen with split virtual drivers (front-end and back-end) arrangement for transparent I/O virtualization. Conditions required for the cooperative cache scheme will not hold for big data as cache hit rate will decrease and disk access will increase due to increased data volume, variety, and variability, resulting in poor performance.

3.1.1.3 I/O targeted approaches

Disk I/O contention is also a problem for the single server virtualization. One solution is to force the disk I/O scheduling on VM and Virtual Machine Monitor (VMM) levels. The selection of optimal disk I/O scheduler for collocated VMs can significantly improve disk I/O resources [41]. However, the tradeoff between disk I/O fairness and performance of collocated VMs is not very flexible and can be achieved at the cost of I/O throughput and latency. Big data either require multiple disks or separate NAS devices for storage. In both cases I/O scheduling at multiple levels is required to avoid disk I/O contention as a whole. So, the disk I/O scheduling only at VM level or VMM level will not work for big data.

A priority based scheduling technique has been proposed in Ref. [42] that improves the response time of I/O requests. In order to improve the response time an I/O request is interrupted at VMM level and the requests are reordered on physical server in the disk I/O queue according to priority. The network bandwidth problem can be solved through hardware solutions such as attaching multiple network adapters or using multi-queue based network adapters. It can also be solved through software solutions such as bandwidth capping, where a bandwidth cap is set in the VM configuration before booting [34]. For big data having increased number of I/O requests two problems will occur: First, a very long queue will be required and second, the priority queue's efficiency decreases with the increase in length as less priority tasks face longer delays.

In Nicpic [43], the state management packet scheduling task responsibility has been separated from CPU. The tasks such as classification of packets, putting them in queue, and application of rate limits are performed by CPU. The packets are queued in memory and scheduled for extraction according to the rate limit of VMs. Both scheduling and extraction tasks are performed by Network Identification Card (NIC) and direct memory access mode is used to extract the packets from queue to NIC. A very long queue is required to handle big data resulting in more memory requirement at NIC level. Since NICs are not capable of handling complex computations, the scheduling and extraction of large number of data packets will become difficult and time consuming resulting into delayed services.

A packet aggregation based mechanism overcomes the memory latency problem caused by network I/O while transferring packets from driver domain to guest OS [44]. Driver domain is used to provide access to shared network devices. One large packet is created by combining several small packets and transferred to the aggregation destination. More data can be transferred with aggregation mechanism since there will be less number of memory requests and revocations, fewer copies, and reduced notifications. Due to big data, the frequency of requests to driver domain at aggregation destination will increase. The reversal of the aggregation process requires extraction of small packets incurring more CPU load, hence resulting in overall performance degradation.

3.1.1.4 Memory targeted approaches

A Xen balloon driver [45] based framework has been proposed to control memory transaction upon the joining of multiple VMs [46]. This framework has been implementation in user space to avoid virtual machine monitor interference. A global–scheduling algorithm solves linear equations, achieves a global solution that adjusts itself according to the availability of resources using dynamic baselines. A working model of the discussed framework has been shown in Fig. 7. In the presence of big data where

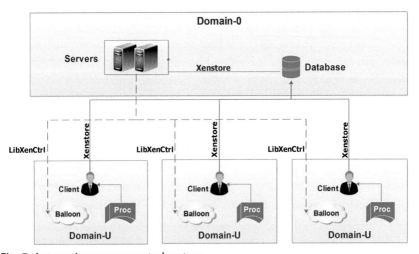

Fig. 7 Automatic memory control system.

multiple VMs, distributed algorithms, and storage drives are employed for a single larger task, the existence of a global-scheduling algorithm is difficult to realize.

3.1.2 VM assignment and placement optimization

The second approach to solve the problem of VM performance overhead is to find an optimal mapping between VMs and physical servers. The techniques presented in Refs. [47,48] propose a prediction based algorithm using a combination of weighted mean with the linear and non-linear model, respectively. Statistical machine learning has been used for reasoning and discovery of optimal mapping. Based on the predicted data, an interference aware scheduler is employed that ultimately reduces performance interference. However, due to high volume, velocity, variety, variability, and veracity of the data, an accurate prediction is challenging.

A cache cloning scheme has been introduced in Ref. [49] for better VM placement. The cache usage estimation is performed through active probing and used for the prediction of performance degradation of an application when placed with other applications. A synthetic cache loader benchmark is used to create a clone for every application arriving at the host platform. The purpose of an application's cache clone is to mimic its cache pressure, which includes both the cache space and the memory bandwidth occupancy of the application. The cache clones are later used as a proxy for actual applications while predicting performance degradation in co-location scenarios. As already discussed, due to high volume, velocity, variety, variability, and veracity of the data, an accurate prediction is hard to make. On the other hand, probing itself becomes a problem when a large volume of data traffic is to be analyzed.

A mechanism to improve VM performance using pSciMapper framework has been discussed in Ref. [50]. The proposed scheme helps in establishing the requirement of physical servers by assigning workflow and VM to each server. A dimensionality reduction technique using Kernel Canonical Correlation Analysis (KCCA) is used to connect the key features with power consumption and time of execution. Hierarchical clustering and the Nelder-Mead optimization algorithm are used to search for optimal consolidation. However, involvement of big data creates a problem for the discussed approach as increased variability of the workflow directly affects the requirement of physical servers and choice of VM, making it difficult to find an optimal consolidation.

A Last Level Cache (LLC) based approach has been presented in Ref. [51] to overcome the VM resource contention. The VMs on physical servers are tracked down having maximum and minimum LLC miss rate. VM having maximum LLC miss rate is swapped with the VM having minimum LLC miss rate. Since in the presence of big data, the cache change will be very rapid and the frequency of LLC miss will be increased posing a greater overhead to the performance of LLC based approach. However, the threshold calculations for optimal performance are not available to evaluate this approach more accurately for big data.

The approach with qualitative prospect in Ref. [25] to identifies resource contention through a clustering technique of low level metrics called DeepDrive. A metric is built by observing application behavior containing information about performance counters and statistics received from hypervisor. Main components of DeepDrive have been shown in Fig. 8.

DeepDrive uses two methods of interference analysis. The results of accuracy and overhead calculations differ for both methods. A warning system is being deployed on VMM which provides early interference statistics. These interference statistics are placed in a multi-dimensional space where both interference and non-interference cases are placed in separate clusters. However, the processing involved for interference analysis increases with the arrival of big data and behavioral analysis becomes very hard due to the requirement of larger datasets analysis.

Xu et al. proposed a technique in which they first identify latency sensitive applications and transfer them to a well-resourced VM for performance improvement [52]. The focus of this study is on the tail of round-trip latency as it produces diversified impact on user experience. An example of

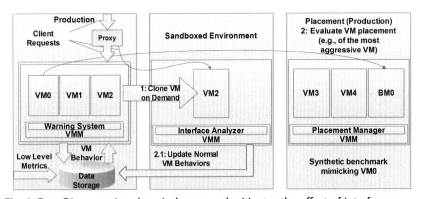

Fig. 8 DeepDrive overview, how it detects and mitigates the effect of interference.

the Amazon's EC2 has been used and the analysis show that the induced latency is due to the poor response time of nodes rather than the network. Applications handling big data are even more resource hungry than latency sensitive applications, but it has not been considered in the proposed scheme. In addition, the element of round-trip latency is not a major problem for data intensive applications.

The discussion about single server techniques show that existing techniques are unable to support big data applications. A comparative analysis of all aforementioned techniques from big data perspective has been presented in Table 3.

3.2 Single data center techniques

The performance of VMs can be degraded due to routine operations such as concurrent deployment, migration, and snapshotting. However, major focus from single data center point of view is on the migration operation as it brings significant overhead [53,54]. Migration in a single data center is performed over LAN. We can further divide VM migration techniques for the single data center in two categories for better understanding: First is based on VM and migration destination selection and second is based on data size and network traffic. We have analyzed VM migration techniques for the single data center from big data point of view.

3.2.1 VM and migration destination selection

Liu et al. [55] has developed a cost model based on performance and energy modeling to calculate the cost of each VM candidate for migration. VM candidate having the least cost is selected for migration. An application-oblivious model has been devised that uses learned knowledge about the workload at VMM level to predict the cost. However, workload analysis of a big data job at VMM level can cause increased number of interferences at VMM level resulting in poor performance. When distributed storage drives are employed, the knowledge at VMM level is not enough to accurately calculate the cost of a VM.

The scheme in Ref. [56] observes the VM migration effects on three entities, i.e., migrating VM, collocated VMs, and applications running on them. The migration operation consumes a lot of CPU cycles and hence affects the performance of both source and destination servers. The concept of Service Level Objective (SLO) penalty is a key consideration of the proposed approach. The VM with the smallest SLO penalty is selected for migration. Since SLO penalty is measured offline at VM level, the

Table 3 Comparison of approaches to improve VM performance overhead for single server.

Techniques	Performance overhead parameters										
	CPU	CPU cache	Memory	Memory B/W	Disk space	Disk I/O	Disk cache/Buffer I/O	Network I/O	Network B/W	Control bus	Data bus
CPU core compensation [38]	✓	✓	–	–	–	✓	✓	✓	–	–	–
Hot page coloring [39]	–	✓	✓	–	–	–	–	–	–	–	–
VMM- and VM-Level I/O schedulers [41]	–	–	–	–	–	✓	–	–	–	–	–
Priority-base I/O schedulers [42]	✓	–	–	–	–	–	–	✓	–	–	–
Rate limiting [43]	–	–	–	–	–	–	✓	–	–	–	–
Copy based cache service [40]	–	–	–	–	–	–	–	✓	–	–	–
Packet aggregation-based mechanism [44]	–	–	✓	–	–	–	–	–	–	–	–
Automatic memory control [46]	✓	–	✓	–	–	–	–	–	–	–	–
Interference modeling [47]	✓	–	–	–	–	✓	✓	–	–	✓	–
Interference calculation [48]	–	–	–	–	–	✓	✓	–	–	✓	–
Cache clone [49]	–	–	–	–	–	–	–	–	–	–	–
Metrics correlation [50]	✓	–	✓	✓	–	✓	✓	–	–	–	–
Cache partitioning [51]	✓	✓	✓	✓	–	–	–	✓	✓	–	–
VM assignment [52]	–	✓	✓	–	–	–	–	–	–	–	–

NAS placed outside the VM is ignored thus rendering the SLO measure inaccurate. Big data storage drives attached to VMs must be considered while measuring SLO penalty. Neglect of the element of storage renders the proposed approach unsuitable for big data applications.

The idea based on a combination of Ant Colony Optimization (ACO) and 2-opt local search algorithms optimally places the VM. This reduces the maximum link utilization [57]. The traffic statistics are obtained from a hypervisor and proposed algorithms are applied for obtaining the optimal candidate to place the VM. However, only traffic statistics at hypervisor level are not enough for big data handling and management.

iAware [58] uses an analytical model to calculate the effects of VM migration. The VM with least effecting factor is chosen for migration. The calculation for the least effecting VM is performed online using a multi-source demand and supply model. Only the effects of VM are calculated and the decision is solely based on this calculation. Big data requires that the impact of other parameters such as volume, velocity, variety, and veracity are considered.

3.2.2 Data size and network traffic optimization

A technique based on pipeline parallelism has been proposed in Ref. [59]. The proposed technique tries to balance out both the processes and data required for migration operation by using the idle CPU and network resources. This helps to expedite the parallelization opportunities available for the live VM migration. However, the scope of the proposed technique is limited to VM level. Distributed data storage drives and impact of big data challenges has not been addressed.

There are many other approaches to overcome data size and network traffic problem, one such example is transfer of data during VM migration by reducing its size. This problem can be mitigated by using any data compression technique like MECOM [60] but will definitely require more CPU cycles. Since CPU is not considered to be a major problem in VM performance, so, one can get the benefit by shifting the network load to CPU. But how much load should be shifted is still a question? The same compression algorithm should be implemented on the destination server to decompress the compressed contents and restore memory state from it. The majority of compression based techniques are only applied to VM and memory. VM and memory are very less in size compared to the affiliated big data storage drives. Both compression and decompression tasks consume large amount of CPU cycles, and for big data the compression and decompression operations can cause CPU resource contention and delay due to the processing of large volume of data. Hence, the proposed techniques to reduce data size and network traffic are inadequate for migrating big data storage and applications.

There are some techniques that reduce the network traffic during live VM migration. Each technique uses a different method to achieve the same goal (e.g., [61–64]) of reducing network traffic. Delta Compression Technique (DCT) also reduces the network traffic by transferring only the changed data between the current and previously transferred data instead of transferring the whole image or disk block in each round of pre-copy. DCT is particularly suited for an application having more memory dirty rate and slow network connection. Compression and decompression operations consume a lot of CPU cycles and cause delay when a large volume of data is used. Further, identification of changed data from a very large dataset is also a CPU intensive task adding more intensiveness to the CPU resource.

On the other hand, the memory pruning [61] or de-duplication method [62,63] transfers only the critical data of OS kernel and application, which is necessary to run the VM and the application. The workflow of Generalized Memory De-duplication (GMD) engine has two stages. The Introspection stage identifies free memory pages in guest VMs and the de-duplication stage duplicates pages using hashing and byte-by-byte comparison. However, hashing and byte-by-byte comparison for the big data requires more CPU cycles and time. If a big data application is running on the VM in migration process, handling the critical data of such an application is not a problem rather than handling and management of the associated big data flows.

CR/RT-Motion [64] uses a log based scheme. It logs events, and when migration is required, these logs are only transferred to the destination server and replayed to restore the memory state. Logging each and every event for a large volume of data may consume more CPU cycles. The concept has not been tested from the big data point of view. However, it will require more computing resources.

In Jo et al.'s technique [65], only those memory pages are transferred that are not available on shared storage. The information of memory location and list of storage blocks is also transferred to the destination server. This information is used to restore the memory state of the migrated VM on the destination server. If we consider the same mechanism for big data storage drives, we can observe that finding available contents and their memory locations from a large volume of data will be difficult to handle. Further, compiling the list and transferring storage blocks for such large and rapidly changing dataset will have an impact on CPU and bandwidth consumption which has not been considered in discussed technique.

Some more schemes focusing on the reduction of network traffic include live gang migration [29]. The live gang migration scheme proposes that same operating systems, applications, and libraries can have a major amount of similar memory contents. Hence, the objective is to locate similar memory contents and transfer a single copy resulting in a reduced amount of network traffic and data. The method of de-duplication [62,63] technique is widely used to find the same memory contents in VM memory and disk image. Afterward, only a single copy of the contents is transferred. The concept of finding similar contents is a successful factor for VM memory, but there is a less chance to find the similar contents on storage. Hence, storage migration is still an unaddressed challenge.

Breitgand et al. [66] has introduced a mathematical model to learn the tradeoff between the VM performance and migration time. The results obtained from the mathematical model is used to allocate a suitable amount of bandwidth for live migration. When same number of VMs are considered for migration and two different techniques are used, it has been observed that sequential transfer of the VMs causes less overhead than a concurrent live migration [67]. The scope of the proposed model is limited to VM only. Affiliated storage and its migration has not been considered.

The significance of the Named Data Networking (NDN) technique [68] is that it uses names of the services to identify the communication with a VM, while other techniques use network addresses. Routing is also performed via service names which results in reduced migration and downtime, and also achieve location independent environment for services and VM. The proposed technique can be employed for big data applications, but no testing has been performed from big data perspective. A comparative analysis of all aforementioned techniques from big data perspective has been provided in Table 4.

3.3 Multiple geo-distributed data centers techniques from big data perspective

The VM migration techniques used for geo-distributed data centers differ from the single data center due to the involvement of WAN. Similar to the single data center platform, we divide geo-distributed data center approaches in two categories: First is based on VM and migration destination selection and second is based on data size and network traffic. However, more focus is on the latter category as in geo-distributed data centers, greater

Table 4 Comparison of approaches to improve VM performance overhead for single data center.

Techniques	Performance overhead parameters							
	Migration time	Migration B/W	Migration traffic	Migration size	Migration speeding	Migration cost	Application down time	Big data ready
Block comparison [29]	✓	–	–	–	–	–	–	–
Cost metrics [55]	✓	–	–	–	–	–	–	–
SLO penalty [56]	–	–	–	–	–	✓	–	–
VMP scheme [57]	✓	–	–	✓	✓	–	–	–
Least performance interference [58]	–	–	–	–	–	✓	–	–
Parallelizing migration [59]	–	–	–	–	✓	–	–	–
Compression [60]	✓	–	–	–	–	–	–	–
Filter useless memory pages [61]	✓	–	–	✓	–	–	–	–
Compression [62]	–	–	✓	–	–	–	–	–
De-duplication [63]	✓	–	–	–	–	–	–	–
Transfer and replay of execution log [64]	✓	–	–	–	–	–	–	–
Transfer location of memory [65]	✓	–	–	–	–	–	–	–
Mathematical modeling [66]	–	✓	–	–	✓	–	–	–
Sequential transfer [67]	–	–	✓	–	–	–	–	–
Named data networking [67]	✓	–	–	–	–	–	✓	–

challenges are limited bandwidth and latency over WAN. Especially when big data is employed, the challenges of limited bandwidth and latency become more severe.

3.3.1 VM and migration destination selection

The technique proposed in Ref. [69] addresses the correlated VM problem which is due to the distribution of multi-tier application over different VMs. All correlated VMs should be transferred to the new destination for proper functioning of the multi-tiered application. A coordinated system called VM buddies has been devised that uses a synchronization protocol to ensure the concurrent migration of all correlated VMs. An adaptive network bandwidth allocation algorithm and a costing model based on pre-copying is used. However, the proposed technique produces a large volume of data in the form of VM disk images and storage drives. Big data in the form of distributed storage drives must also be transferred along with VM disk images, but the proposed technique lacks in providing a solution for storage migration. If relevant big data is not transferred and a link to distributed storage drives is made, then network latency over WAN effects the performance.

Another solution for the problem of correlated VM problem has been presented in Ref. [70]. Following condition is imposed in the proposed solution: Correlated VMs should belong to a single Virtual Data Center (VDC) and to a single VDC request. However, the request may consist of virtual resources from multiple VDCs.

3.3.2 Data size and network traffic optimization

The first attempt to transfer a running web server over a WAN environment has been made by Bradford et al. [36]. In order to complete the migration process, an iterative loop was created through which the memory state and disk image of the VM was transferred. There is a bottleneck in the proposed migration process, when write intensive workloads are to be transferred, loop iterations are drastically increased. However, the problem of increased iterations has been addressed through write throttling mechanism that limits the number of write operations to control the transfer and disk dirty rate. The current network connections are dynamically redirected to the destination server using IP tunneling technique. However, the proposed technique does not cover the migration of affiliated storage drives incorporating big data. Incase big data is considered; the write throttling technique will create a bottleneck due to increased number of write requests. This results in a very long iterative loop which not only consumes more CPU cycles but network and bandwidth resources as well. The complete process will also take more time than expected, which is not desired.

The mechanism in Ref. [36] reduces the number of dirty pages and rounds of memory pre-copy phase by transferring only the difference between current and previously transferred pages. The cost of VM migration is higher than the memory state transfer. The migration cost is reduced by replicating the VM images in the background before live migration, and then synchronizing the disk states later [71]. However, comparison of current and previously transferred pages involving big data will take more time due to the increased number of comparisons. The incremental effect in comparisons will also bring a heavy load on CPU effecting the performance and making the proposed approach unsuitable for big data applications.

Another technique proposed by Akiyama et al. [71] is for the data intensive applications that uses page cache teleportation for the reduction of data size to be transferred. The proposed technique uses memory addresses to successfully migrate the VM to the destination server and the process of memory location identification is performed before the actual migration is in the process. The whole operation is performed using the following steps: (1) The disk image is synchronized before a migration. (2) Write access to the memory are tracked during the following steps. (3) The kernel module detects where the restorable page cache is on the guest memory and the disk. (4) The module sends them to the VMMs. (5) The destination VMM copies the disk blocks to the guest memory. (6) The source VMM sends memory pages except the restorable page cache, which has not been modified during the steps (3)–(5). Storage migration is not covered in the proposed technique. However, if we analyze the proposed technique from big data storage perspective, we can say that larger storage size will take more time to synchronize. Tracking a large number of write operations and locating unmodified and restorable pages will definitely consume more CPU cycles making the proposed technique resource intensive, costly, and undesirable.

A prioritization based technique has been proposed in Ref. [72] to transfer VM data. The proposed method prioritizes the transfer of critical data that are the data required for booting the OS and running the user application. However, the proposed technique lacks in providing the solution for the non-critical data. For a big data application, critical data are not a problem, but big data flow is a major challenge that has not been considered in the proposed technique.

An I/O mirroring technique based on VMWare ESX server has been developed by Mashtizadeh et al. [73]. It works by blocking the read and write I/O operations on the source server until the VM image is successfully migrated on the destination server. The frequency of big data read and write operations is very high and holding such operations for a small amount of time will create a very lengthy queue. The proposed technique lacks in providing any solution for how to safely block such a large number of read and write operations. However, blocking such large number of read and write operations will result into delayed services that is not desirable for customers and CSPs.

Another technique that uses a combination of on–demand copying and background copying concepts to transfer a disk image rapidly over the WAN having less impact on I/O performance [74]. The proposed solution is implemented as a proxy server using storage I/O protocol (e.g., iSCSI). This solution can be integrated into network attached storage devices available in a data center, hence making it independent of VMM implementations. This technique is indeed promising and can be employed for big data applications, but no testing has been performed from this perspective. Also, other WAN parameters such as latency and bandwidth must be considered and calculations regarding transfer time needs to be performed.

However, all aforementioned approaches induce heavy I/O latencies, especially in the case of I/O intensive workload migration. Nicolae and Cappello [75] have further minimized I/O latencies by reducing data size and network traffic. First one transfers only the modified content excluding the VM's OS. Second technique is more complex as it involves a hybrid data fetching method in which push and pull methods are used. Cold data are pushed by the source server and the dirty data are pulled by the destination server. The server possessing the transfer-control is only allowed to perform the operation. When we discuss about big data, the time to locate and identify modified contents is very high due to a large volume of data. The process of locating modified contents will consume more CPU cycles. The processing of cold and dirty data is also a CPU intensive task. In addition, the server not possessing the transfer control will have to wait for longer time due to the processing of other server on big data holding transfer-control.

A scheduling technique has been discussed in Ref. [76] that is workload-aware and based on disk block transfer. In the proposed technique, disk blocks are scheduled to be transferred in an optimized way resulting in minimized network traffic and improved I/O performance. The concepts of spatial locality, temporal locality, and popularity characteristics of read, write, and disk block operations are used as key factors to schedule ordering of data blocks. However, the proposed technique consumes more CPU cycles. Mostly CPU is not considered a major limiting factor in geo-distributed migration, but when big data disk blocks are to be transferred the CPU might become a limiting factor. The reason is that the calculations of spatial locality, temporal locality, and popularity for such a large amount of read, write, and disk block operations will consume a much greater number of CPU Cycles.

In Ref. [77], Jadhav has proposed a scheme to perform a VM disk migration based on the principal of I/O mirroring. An iterative copy approach is used for VM memory state migration. A point-to-point connection between source and destination server is established, and synchronous mirroring technique is used to transfer the VM disk. For big data the number of iterations for copying such a large volume of data will be increased and synchronous mirroring will also take longer time. The chances of errors in the synchronization process will also be increased as link stability over WAN is lesser than in a LAN environment.

From discussion about geo-distributed data center techniques, we can conclude that existing techniques are not big data ready. A comparative analysis of all aforementioned techniques from big data perspective has been provided in Table 5.

4. Summary, insights and open research challenges

A possible future cloud model may consist of SDN and big data that are main ingredients of future Internet [78]. The concept of cloud federation has already been coined, in which CSPs collaborate with each other for resource sharing and load balancing [79]. SDNs can play a pivotal role in the processing of big data due to features such as programmability, control, and efficiency of data transfer [80]. A possible consolidated model of the future cloud has been provided in Fig. 9.

Every CSP in the federation will have a control plane and data plane. Programmability is employed to cope with the volatile network change and reduce cost of the equipment. The control plane will work in a distributed manner connecting to all other control planes. Since control plane is responsible for control traffic that has a low data rate and can be managed easily, higher data rates can be achieved at data plane using traffic flow rules to aggregate and partition traffic flows according to link condition [18]. When a certain migration is triggered, the connections to the affiliated NAS can be transferred temporarily and data shifting process from source NAS to destination can be scheduled in off-peak hours. After the successful

Table 5 Comparison of approaches to improve VM performance overhead for geo-distributed data centers.

Techniques	Performance overhead parameters							
	Migration time	Migration B/W	Migration traffic	Migration size	Migration speedup	Migration cost	Application down time	Big data ready
Write throttling IP tunneling [35]	✓	–	–	–	–	–	✓	–
Pre-copy [36]	–	–	✓	–	–	✓	✓	–
Correlated VM migration [69]	✓	–	–	–	–	–	✓	–
Virtual data center migration [70]	✓	–	–	–	–	–	✓	–
Page cache restoring [71]	–	–	–	–	✓	–	–	–
De-duplication, data priority transfer [72]	–	–	–	✓	–	–	✓	–
I/O mirroring [73]	✓	–	–	–	–	–	–	–
On-demand fetching, background copying [74]	–	–	–	–	✓	–	–	–
Modified data transfer, data pre-fetching [75]	–	–	✓	–	–	–	–	–
Data block scheduling on locality, popularity [76]	✓	–	–	–	–	–	✓	–
I/O mirroring [77]	✓	–	✓	–	–	–	✓	–

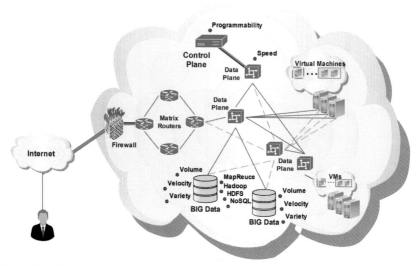

Fig. 9 Possible future cloud model.

migration of data, the connections to old data can be terminated. A grace period must be set in order to make sure that the migrated VM and data is successfully transferred to the destination site and work as required. After the expiration of that grace period, the respective VM and data can be removed.

An amalgamated view of the VM performance parameters has been presented in Fig. 10 along with the summary and future directions keeping big data in perspective. It is evident from the overall discussion in this chapter that much work has been done in the area of shared resource management and LAN migration. However, areas such as WAN migration, downtime, and especially big data storage migration are still open research issues.

4.1 VM migration and big data

VM migration techniques serve as a base for managing computing resources, minimizing VM performance overhead, achieving energy efficiency and load balancing in cloud computing [77,81].

Currently, the aforementioned VM migration techniques lack in handling the velocity, variety, and variability of the data. In WAN, the bandwidth requirements for such—data intensive applications are very high posing another challenge for the low bandwidth networks. The network

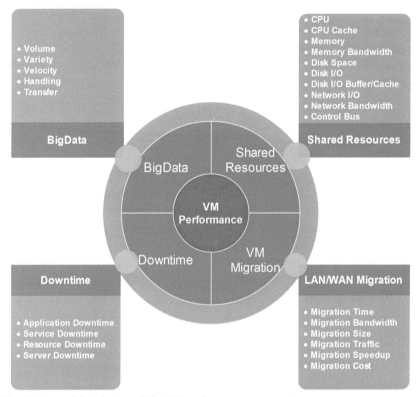

Fig. 10 Consolidated view of the VM performance parameters.

infrastructure also lacks in accommodating such large sized requests. On a low bandwidth network, aforementioned techniques will perform poorly due to the delayed migration as frequent updates on the source image will generate more network traffic. However, making the VM design less dependent upon applications can solve the problem. For example, for a web server if only PHP and MYSQL is required to run on the server then we can only move the application and its data instead of migrating the whole VM in a standard web server VM image.

4.2 VM storage migration and big data

VM migration techniques manage the performance overhead in a cost effective manner. When discussing about performance overhead management, a VM migration technique emphasis on two methods through which the desired results can be achieved. One is to reduce the size of data to be transferred during the VM migration process between source and destination server and second is to reduce the amount of memory requirement or disk dirty rate.

Techniques in Refs. [36,60–63,72] fall in the category of minimizing the size of data transferred or disk dirty rate. What about the storage migration which is a vital part of the migration process? Big data storage drives are difficult to migrate along with VM images generating more data and network traffic. Second, the techniques falling in the category of minimizing the size of data transferred or disk dirty rate achieve their goal by blocking I/O requests to the migrating VM, while in the process of migration (e.g., [72,73]). In case of big data, where the number of write requests is very high, there should be a large enough buffer to temporarily hold the incoming requests. Otherwise, the users will face downtime of services. However, if the option of buffering such larger requests is considered, the load will definitely shift to computing resources such as CPU and memory effecting the collocated VMs and applications running on them. Hence, the design of such an approach that will cover the objective of consuming less computing resources without affecting the collocated VMs is still a challenge. If the need of storage migration can be eliminated in a way that storage can be accessed from anywhere, then the relevant data can easily be assigned to the newly migrated VM. But the solution for assigning data to VM will not work for a data intensive application due to communication latency introduced by the newly assigned route and channel over WAN.

4.3 Effect of VM migration on collocated VMs with big data

The extensive study about VM migration techniques shows that the majority of the techniques are either focused on the performance overhead caused by VM migration or on the effects of migrated VM on source and destination servers (e.g., [50,56]). There are not many techniques that have discussed the effects on VM collocation on a single destination server due to live migration [58]. As the VM is transferred to a new physical server, it fights for the limited physical resources, causing an effect on the collocated VMs hosted on the destination server. However, this is true for both source and destination servers as for time being during migration, additional resources are required on the source side, and after migration, the resources are released providing more room for the collocated VMs on the source server.

Big data storage migration is more challenging as many distributed algorithms work on storage and databases. It becomes hard to establish a global picture in the presence of distributed algorithms as the frequency of data change and WAN latency is very high. It is very critical to transfer the relevant information, schema, and data with the VM. It is required by the VM performance overhead mitigation techniques to consider both factors causing the overhead, VM performance overhead due to migration of VM and storage, and its effects on collocated VMs on source and destination servers. Establishing such a performance centric approach can be a challenging research problem. However, by improving VM resource prediction techniques for destination server, it is possible to provision required resources in advance, thus avoiding migration effects on collocated VMs at run time.

4.4 Cost effectiveness and big data

In cloud data centers, finding a tradeoff between the monetary cost of hardware solution (i.e., network switches, servers, storage drives, and network links) and VM performance overhead is an appealing research area. With the increase of data and network traffic, it has become inevitable to work with low bandwidth and simple hardware. The management and processing of big data requires complex and speedy hardware. A simple solution could be rendering more hardware for the same number of VMs. It will definitely help in improving performance but will increase the cost which is not a desirable factor for the cloud service providers. In an extreme case, it is possible to create a one-to-one mapping solution, that is one VM to a single physical server, one logical link to single physical connection, one VM storage to a physical storage, but the resulting solution will become highly impractical. However, in this example the resource utilization will be very low and VM performance overhead is either at minimal level or removed completely which is against the interest of the cloud service providers. Hence, this research is very vital to evaluate the tradeoffs between required physical resources, their utilization, and sharing, keeping the performance overhead factor and big data challenges in perspective. A combined effort in all directions is required to produce cost effective solutions using less sophisticated and low computing hardware requiring less bandwidth and network resources.

4.5 Network traffic, heterogeneity and big data

Majority of the network sharing solutions are those that are based on the hose model (e.g., [82–86]. In hose model, dedicated network connections are used to connect all VMs with a non-blocking logical switch. Network capacity of the central switches is not infinite and hose model only provides simplification for the VM's networking paradigm. Network congestion cannot be avoided completely whether a small or large data center network with the enormous amount of bandwidth [87] is considered. Network congestion is the main reason due to which hose model is impractical in real world data centers. Hose model also does not provide any means to manage the greater velocity of the data over the network. The question still remains, that how a network model should be designed to facilitate the network sharing solution keeping the heterogeneity (e.g., [86,88]) and big data challenges in perspective.

The majority of the existing VM allocation techniques have targeted homogenous network formation and bandwidth requirement, few solutions have studied the heterogeneous nature of network and bandwidth requirement which further complicates the bandwidth sharing problem. More focus and research is required to develop VM and network sharing solutions targeting the heterogeneous nature of the network and bandwidth requirement on physical as well as logical level. In a cloud data center, the hardware heterogeneity is possible [89], but even within the same type of Amazon EC2 instances, it does not mean that it is also ready for big data. Accommodating the heterogeneity factor along with big data challenges while applying techniques for VM migration and performance overhead mitigation can be challenging. However, if an intelligent control system can be developed using the programmability feature of devices on the fly as offered by SDNs, the adaptability to work with different networks having different data and network requirements can become easy. The adaptability feature will help in balancing resources according to their unique requirements efficiently and the problem of heterogeneity can be resolved to some extent.

Need for computing resources is highly volatile and this fact has also been proven by Google's production data center analysis [90]. Different kinds of workloads running simultaneously in the data center makes the resource demand prediction more difficult. Another fact is that the majority of the VM performance management solutions work on the principle of accurate resource prediction, which is very difficult when data variety, velocity, and variability is very high. Accurate resource prediction in heterogeneous and volatile workload with less prediction error is very challenging.

4.6 Hybrid of public and private clouds with big data

It is an observation that the public and private clouds are used in combination forming a hybrid model to facilitate the customers. Customers may jump between public and private clouds as per their need of resources and are charged accordingly [87]. This private-public hybridization is also used to accommodate the needs produced by big data applications. However, customers are not ready for the related challenges of data capture, data storage, data analysis, data visualization, required bandwidth, velocity, variety, and variability. There is a strong need for designing new applications by keeping the hybridization factor in perspective.

If we take the example of "cloud bursting" [91], where synchronization becomes more critical as the service runs in the local data center and its replica runs on the public cloud. Synchronization of big data will be very difficult as it requires synchronization of VM's state and the snapshot across the network, thus producing a large volume of data to be transferred between private and public cloud replicas. Are the existing solutions ready to support such heavy and inflexible resource demand? These are all open questions for the future research to address keeping the VM performance and big data in perspective, especially across multiple geo-distributed data centers. However, the problem of synchronization can be minimized if we can develop such scheme for the deployment of VMs having a combination of data intensive and non-intensive applications running on different VMs, but on the same physical server. So, if there is a need for more resources by a VM running data intensive application, the VM having less data intensive application can be migrated instead. Hence, local resources can be provided at priority to the VM having data intensive applications and the other VM having less data intensive application can use Cloud Bursting resulting in less data for synchronization.

5. Conclusion

This chapter presents VM performance issues, effecting factors, and overhead of mitigation techniques. Resource parameters and challenges related to different operations and big data have been established to assess overhead mitigation techniques. It has been concluded that existing VM migration techniques are not ready to handle big data, especially the aspects of big data application and storage migration are yet to be incorporated. A possible future cloud model has been discussed that will help in developing new techniques to address big data challenges. The future cloud model consists of SDN and cloud federation to support big data. The programmability feature of SDN provided by control plane can be leveraged to attain customized underlying network configurations suitable for big data. Moreover, the data plane of SDN helps in providing sufficient data rate due to separation of control decisions, thus making it suitable for big data applications and storage migration.

References

[1] C.P. Chen, C.Y. Zhang, Data-intensive applications, challenges, techniques and technologies: A survey on Big Data, Inform. Sci. 275 (2014) 314–347.

[2] IDC Research Inc., Online available at, http://www.idc.com/getdoc.jsp?containerId=prUS25946315. Accessed (2016).

[3] F. Xu, F. Liu, L. Liu, H. Jin, B. Li, B. Li, Iaware: making live migration of virtual machines interference-aware in the cloud, IEEE Trans. Comput. 63 (12) (2014) 3012–3025.

[4] H. Liu, H. Jin, C.Z. Xu, X. Liao, Performance and energy modeling for live migration of virtual machines, Clust. Comput. 16 (2) (2013) 249–264.

[5] G.V. Asokan, V. Asokan, Leveraging "big data" to enhance the effectiveness of "one health" in an era of health informatics, J. Epidemiol. Glob. Health 5 (4) (2015) 311–314.

[6] R. Mall, R. Langone, J.A. Suykens, Self-tuned kernel spectral clustering for large scale networks, in: Big data, 2013 IEEE International Conference, IEEE, 2013, pp. 385–393.

[7] R. Branch, H. Tjeerdsma, C. Wilson, R. Hurley, S. McConnell, Cloud computing and big data: a review of current service models and hardware perspectives, J. Softw. Eng. Appl. 7 (8) (2014) 686.

[8] R. Mall, R. Langone, J.A. Suykens, Kernel spectral clustering for big data networks, Entropy 15 (5) (2013) 1567–1586.

[9] L. Borovick, R.L. Villars, The Critical Role of the Network in Big Data Applications, Cisco White Paper, 2012.

[10] P. Chopade, J. Zhan, K. Roy, K. Flurchick, Real-time large-scale big data networks analytics and visualization architecture, in: Emerging Technologies for a Smarter World (CEWIT), 12th International Conference & Expo, IEEE, 2015, pp. 1–6.

[11] A.A. Hamed, X. Wu, T. Fandy, Mining patterns in big data kh networks, in: IEEE/ACS 11th International Conference on Computer Systems and Applications (AICCSA), IEEE, 2014, pp. 176–183.

[12] R. Mall, V. Jumutc, R. Langone, J.A. Suykens, Representative subsets for big data learning using k-NN graphs, in: Big Data (Big Data), IEEE International Conference, IEEE, 2014, pp. 37–42.

[13] R. Mall, R. Langone, J.A. Suykens, Multilevel hierarchical kernel spectral clustering for real-life large scale complex networks, PloS One 9 (6) (2014) e99966.

[14] Introduction to Big data: Infrastructure and Networking Considerations Leveraging Hadoop-Based Big data architectures for a scalable, High-Performance analytics Platform, Juniper 2012. Accessed (2016).

[15] D. Mysore, S. Khupat, S. Jain, Big Data Architecture and Patterns, Part 3: Understanding the Architectural Layers of a Big Data Solution, October 2013 Accessed. IBM developerWorks, 2016.

[16] L. Borovick, R.L. Villars, The Critical Role of the Network in Big Data Applications, Cisco White paper, 2012.

[17] P. Heller, D. Piziak, J. Knudse, An Enterprise Architect's Guide to Big Data— Reference Architecture Overview, Oracle Enterprise Architecture White Paper, Unpublished paper, 2015.

[18] G. Wang, T.S. Ng, A. Shaikh, Programming your network at run-time for big data applications, in: Proceedings of the First Workshop on Hot Topics in Software Defined Networks, ACM, 2012, August, pp. 103–108.

[19] B. Tian, Y. Tian, B. Huebert, Y. Sun, T. Hurt, W. Ho, … H. Lee, SecHDFS: a secure data allocation scheme for heterogenous hadoop systems, in: Networking, Architecture and Storage (NAS), IEEE International Conference, IEEE, 2016, pp. 1–2.

[20] Chethan Kumar, Managing Performance Variance of Applications Using Storage I/O Control, VMware, Inc. Palo Alto, CA. 2010. www.vmware.com Accessed (2016).

[21] B. Posey, Virtualization, The Do's and Don'ts of Server Virtualization, Techtarget, 2010, Online available at: http://searchservervirtualization.techtarget.com/tip/Virtualizing-applications-and-servers-Dos-and-donts.

[22] Antivirus Best Practices for VMware® Horizon View™ 5.x, VMware, Inc. Palo Alto, CA. Online available at: https://www.vmware.com/files/pdf/VMware-View-AntiVirusPractices-TN-EN.pdf, Accessed (2015).

[23] Virtualization and Disk Performance, Diskeeper Corporation, Burbank, CA, Online available at: www.condusiv.com/188, 2006. Accessed (2016).

[24] Performance Best Practices for VMware Workstation, VMware, Inc., 2009, Online available at: www.vmware.com/pdf/ws7_performance.pdf, Accessed (2015).

[25] D. Novakovic, N. Vasic, S. Novakovic, D. Kostic, R. Bianchini, Deepdive: Transparently Identifying and Managing Performance Interference in Virtualized Environments, 2013 (No. EPFL-report-183449).

[26] A. Masood, M. Sharif, M. Yasmin, M. Raza, Virtualization tools and techniques: survey, Nepal J. Sci. Technol. 15 (2) (2015) 141–150.

[27] T. Hwang, K. Kim, C. Sung, LACS: latency-aware credit scheduler to improve responsiveness, in: Proceedings of the 2015 Conference on Research in Adaptive and Convergent Systems, ACM, 2015, pp. 474–475.

[28] I.A.T. Hashem, I. Yaqoob, N.B. Anuar, S. Mokhtar, A. Gani, S.U. Khan, The rise of "big data" on cloud computing: review and open research issues, Inf. Syst. 47 (2015) 98–115.

[29] H. Liu, B. He, VMbuddies: coordinating live migration of multi-tier applications in cloud environments, IEEE Trans. Parallel Distrib. Syst. 26 (4) (2015) 1192–1205.

[30] U. Deshpande, X. Wang, K. Gopalan, Live gang migration of virtual machines, in: Proceedings of the 20th International Symposium on High Performance Distributed Computing, ACM, 2011, pp. 135–146.

[31] B. Nicolae, J. Bresnahan, K. Keahey, G. Antoniu, Going back and forth: efficient multideployment and multisnapshotting on clouds, in: Proceedings of the 20th International Symposium on High Performance Distributed Computing, ACM, 2011, pp. 147–158.

[32] J. Reich, O. Laadan, E. Brosh, A. Sherman, V. Misra, J. Nieh, D. Rubenstein, VMTorrent: scalable P2P virtual machine streaming, in: CoNEXT, vol. 12, 2012, pp. 289–300.

[33] S.H. Lim, J.S. Huh, Y. Kim, C.R. Das, Migration, assignment, and scheduling of jobs in virtualized environment, in: Migration, 40, 2011, p. 45.

[34] C. Clark, Xen Users' Manual, Citrix Systems, Inc., University of Cambridge, U.K., XenSource Inc., IBM Corp., Hewlett-Packard Co., Intel Corp., AMD Inc., 2008. Accessed (2016).

[35] R. Bradford, E. Kotsovinos, A. Feldmann, H. Schiöberg, Live wide-area migration of virtual machines including local persistent state, in: Proceedings of the 3rd International Conference on Virtual Execution Environments, ACM, 2007, June, pp. 169–179.

[36] T. Wood, K.K. Ramakrishnan, P. Shenoy, J. Van der Merwe, CloudNet: dynamic pooling of cloud resources by live WAN migration of virtual machines, ACM SIGPLAN Not. 46 (7) (2011) 121–132. ACM.

[37] A. Stuart, K. Ord, Kendall's Advanced Theory of Statistics, Distribution Theory, Wiley. vol. 1, (2010). https://books.google.com.pk/books?id=pSngnQEACAAJ.

[38] R. Nathuji, A. Kansal, A. Ghaffarkhah, Q-clouds: managing performance interference effects for qos-aware clouds, in: Proceedings of the 5th European Conference on Computer Systems, ACM, 2010, pp. 237–250.

[39] X. Zhang, S. Dwarkadas, K. Shen, Towards practical page coloring-based multicore cache management, in: Proceedings of the 4th ACM European Conference on computer Systems, ACM, 2009, April, pp. 89–102.

[40] H. Kim, H. Jo, J. Lee, Xhive: efficient cooperative caching for virtual machines, IEEE Trans. Comput. 60 (1) (2011) 106–119.

[41] D. Boutcher, A. Chandra, Does virtualization make disk scheduling passé? ACM SIGOPS Oper. Syst. Rev. 44 (1) (2010) 20–24.

[42] F. Blagojevic, C. Guyot, Q. Wang, T. Tsai, R. Mateescu, Z. Bandic, Priority IO Scheduling in the Cloud. HGST Research Paper, http://www.hgst.com, 2013.

[43] S. Radhakrishnan, V. Jeyakumar, A. Kabbani, G. Porter, A. Vahdat, NicPic: scalable and accurate end-host rate limiting, in: Presented as Part of the 5th USENIX Workshop on Hot Topics in Cloud Computing, USENIX, 2013.

[44] M. Bourguiba, K. Haddadou, I. El Korbi, G. Pujolle, Improving network, I/O virtualization for cloud computing, IEEE Trans. Parallel Distrib. Syst. 25 (3) (2014) 673–681.

[45] A. Binu, G.S. Kumar, Virtualization techniques: a methodical review of XEN and KVM, in: International Conference on Advances in Computing and Communications, Springer, Berlin, Heidelberg, 2011, pp. 399–410.

[46] W.Z. Zhang, H.C. Xie, C.H. Hsu, Automatic memory control of multiple virtual machines on a consolidated server, IEEE Trans. Cloud Comput. 5 (1) (2017) 2–14.

[47] Q. Zhu, T. Tung, A performance interference model for managing consolidated workloads in QoS-aware clouds, in: Cloud Computing (CLOUD), IEEE 5th International Conference, IEEE, 2012, pp. 170–179.

[48] R.C. Chiang, H.H. Huang, TRACON: interference-aware scheduling for data-intensive applications in virtualized environments, in: Proceedings of 2011 International Conference for High Performance Computing, Networking, Storage and Analysis, ACM, 2011 (p. 47).

[49] S. Govindan, J. Liu, A. Kansal, A. Sivasubramaniam, Cuanta: quantifying effects of shared on-chip resource interference for consolidated virtual machines, in: Proceedings of the 2nd ACM Symposium on Cloud Computing, ACM, 2011 (p. 22).

[50] Q. Zhu, J. Zhu, G. Agrawal, Power-aware consolidation of scientific workflows in virtualized environments, in: Proceedings of the 2010 ACM/IEEE International Conference for High Performance Computing, Networking, Storage and Analysis, IEEE Computer Society, 2010, pp. 1–12.

[51] J. Ahn, C. Kim, J. Han, Y.R. Choi, J. Huh, Dynamic virtual machine scheduling in clouds for architectural shared resources, in: Proceedings of the USENIX Workshop on Hot Topics in Cloud Computing (HotCloud), 2012.

[52] Y. Xu, Z. Musgrave, B. Noble, M. Bailey, Bobtail: avoiding long tails in the cloud, in: Presented as Part of the 10th {USENIX} Symposium on Networked Systems Design and Implementation, NSDI, 2013, pp. 329–341.

[53] S. Akoush, R. Sohan, A. Rice, A.W. Moore, A. Hopper, Predicting the performance of virtual machine migration, in: Modeling, Analysis & Simulation of Computer and Telecommunication Systems (MASCOTS), IEEE International Symposium, IEEE, 2010, pp. 37–46.

[54] C. Clark, K. Fraser, S. Hand, J.G. Hansen, E. Jul, C. Llmpach, … A. Warfleld, Live migration of virtual machines, in: 2nd USENIX Symposium on Networked Systems. Design and Implementation (NSDI 05), 273286, 2005.

[55] H. Liu, H. Jin, C.Z. Xu, X. Liao, Performance and energy modeling for live migration of virtual machines, Clust. Comput. 16 (2) (2013) 249–264.

[56] Z. Shen, S. Subbiah, X. Gu, J. Wilkes, Cloudscale: elastic resource scaling for multi-tenant cloud systems, in: Proceedings of the 2nd ACM Symposium on Cloud Computing, ACM, 2011 (p. 5).

[57] J.K. Dong, H.B. Wang, Y.Y. Li, S.D. Cheng, Virtual machine placement optimizing to improve network performance in cloud data centers, J. China Univ. Posts Telecommun. 21 (3) (2014) 62–70.

[58] F. Xu, F. Liu, L. Liu, H. Jin, B. Li, B. Li, Iaware: making live migration of virtual machines interference-aware in the cloud, IEEE Trans. Comput. 63 (12) (2014) 3012–3025.

[59] X. Song, J. Shi, R. Liu, J. Yang, H. Chen, Parallelizing live migration of virtual machines, ACM SIGPLAN Not. 48 (7) (2013) 85–96. ACM.

[60] H. Jin, L. Deng, S. Wu, X. Shi, X. Pan, Live virtual machine migration with adaptive, memory compression, in: Cluster Computing and Workshops, CLUSTER'09. IEEE International Conference, IEEE, 2009, pp. 1–10.

[61] A. Koto, H. Yamada, K. Ohmura, K. Kono, Towards unobtrusive vm live migration for cloud computing platforms, in: Proceedings of the Asia-Pacific Workshop on Systems, ACM, 2012 (p. 7).

[62] P. Svärd, B. Hudzia, J. Tordsson, E. Elmroth, Evaluation of delta compression techniques for efficient live migration of large virtual machines, ACM SIGPLAN Not. 46 (7) (2011) 111–120.

[63] J.H. Chiang, H.L. Li, T.C. Chiueh, Introspection-based memory de-duplication and migration, ACM SIGPLAN Not 48 (7) (2013) 51–62. ACM.

[64] H. Liu, H. Jin, X. Liao, L. Hu, C. Yu, Live migration of virtual machine based on full system trace and replay, in: Proceedings of the 18th ACM International Symposium on High Performance Distributed Computing, ACM, 2009, June, pp. 101–110.

[65] C. Jo, E. Gustafsson, J. Son, B. Egger, Efficient live migration of virtual machines using shared storage, ACM SIGPLAN Not. 48 (7) (2013) 41–50. ACM.

[66] D. Breitgand, G. Kutiel, D. Raz, Cost-aware live migration of services in the cloud, in: Proceedings of the 3rd Annual Haifa Experimental Systems Conference, ACM, 2010, p. 11.

[67] A. Shieh, S. Kandula, A.G. Greenberg, C. Kim, B. Saha, Sharing the data center network, in: NSDI, vol. 11, 2011, p. 23.

[68] R. Xie, Y. Wen, X. Jia, H. Xie, Supporting seamless virtual machine migration via named data networking in cloud data center, IEEE Trans. Parallel Distrib. Syst. 26 (12) (2015) 3485–3497.

[69] H. Liu, B. He, VMbuddies: coordinating live migration of multi-tier applications in cloud environments, IEEE Trans. Parallel Distrib. Syst. 26 (4) (2015) 1192–1205.

[70] G. Sun, D. Liao, D. Zhao, Z. Xu, H. Yu, Live migration for multiple correlated virtual machines in cloud-based data centers, IEEE Trans. Serv. Comput. 11 (2) (2018) 279–291.

[71] S. Akiyama, T. Hirofuchi, R. Takano, S. Honiden, Fast wide area live migration with a low overhead through page cache teleportation, in: Cluster, Cloud and Grid Computing (CCGrid), 13th IEEE/ACM International Symposium, IEEE, 2013, pp. 78–82.

[72] S. Al-Kiswany, D. Subhraveti, P. Sarkar, M. Ripeanu, VMFlock: virtual machine co-migration for the cloud, in: Proceedings of the 20th International Symposium on High Performance Distributed Computing, ACM, 2011, pp. 159–170.

[73] A.J. Mashtizadeh, E. Celebi, T. Garfinkel, M. Cai, The design and evolution of live storage migration in VMware ESX, in: USENIX Annual Technical Conference, 2011, pp. 187–200.

[74] T. Hirofuchi, H. Ogawa, H. Nakada, S. Itoh, S. Sekiguchi, A live storage migration mechanism over wan for relocatable virtual machine services on clouds, in: Proceedings of the 2009 9th IEEE/ACM International Symposium on Cluster Computing and the Grid, IEEE Computer Society, 2009, pp. 460–465.

[75] B. Nicolae, F. Cappello, A hybrid local storage transfer scheme for live migration of i/o intensive workloads, in: Proceedings of the 21st International Symposium on High-Performance Parallel and Distributed Computing, ACM, 2012, pp. 85–96.

[76] J. Zheng, T.S.E. Ng, K. Sripanidkulchai, Workload-aware live storage migration for clouds, ACM SIGPLAN Not. 46 (7) (2011) 133–144. ACM.

[77] A.J. Mashtizadeh, M. Cai, G. Tarasuk-Levin, R. Koller, T. Garfinkel, S. Setty, XvMotion: unified virtual machine migration over long distance, in: Proceedings of the 2014 USENIX Annual Technical Conference (USENIX ATC 14), 2014.

[78] H. Yin, Y. Jiang, C. Lin, Y. Luo, Y. Liu, Big data: transforming the design philosophy of future internet, IEEE Netw. 28 (4) (2014) 14–19.

[79] D.G. Kogias, M.G. Xevgenis, C.Z. Patrikakis, Cloud federation and the evolution of cloud computing, Computer 49 (11) (2016) 96–99.

[80] L. Cui, F.R. Yu, Q. Yan, When big data meets software-defined networking: SDN for big data and big data for SDN, IEEE netw. 30 (1) (2016) 58–65.

[81] N. Park, I. Ahmad, D.J. Lilja, Romano: autonomous storage management using performance prediction in multi-tenant data centers, in: Proceedings of the Third ACM Symposium on Cloud Computing, ACM, 2012, October (p. 21).

[82] V. Jeyakumar, M. Alizadeh, D. Mazières, B. Prabhakar, C. Kim, A. Greenberg, EyeQ: practical network performance isolation at the edge, REM 1005 (A1) (2013) A2.

[83] L. Popa, P. Yalagandula, S. Banerjee, J.C. Mogul, Y. Turner, J.R. Santos, Elasticswitch: practical work-conserving bandwidth guarantees for cloud computing, ACM SIGCOMM Comput. Commun. Rev. 43 (4) (2013) 351–362. ACM.

[84] L. Popa, G. Kumar, M. Chowdhury, A. Krishnamurthy, S. Ratnasamy, I. Stoica, FairCloud: sharing the network in cloud computing, in: Proceedings of the ACM SIGCOMM 2012 Conference on Applications, Technologies, Architectures, and Protocols for Computer Communication, ACM, 2012, pp. 187–198.

[85] H. Rodrigues, et al., Gatekeeper: Supporting Bandwidth Guarantees for Multi-tenant Data center Networks, WIOV, 2011.

[86] D. Xie, N. Ding, Y.C. Hu, R. Kompella, The Only Constant is Change: Incorporating Time-Varying Network Reservations in Data Centers, ACM SIGCOMM Comput. Commun. Rev. 42 (4) (2012) 199–210.

[87] V. Jeyakumar, et al., EyeQ: Practical network performance isolation for the multi-tenant cloud, in: Proceedings of the 4th USENIX Conference on Hot Topics in Cloud Computing, USENIX Association, 2012.

[88] J. Zhu, D. Li, J. Wu, H. Liu, Y. Zhang, J. Zhang, Towards bandwidth guarantee in multi-tenancy cloud computing networks, in: Network Protocols (ICNP), 20th IEEE International Conference, IEEE, 2012, pp. 1–10.

[89] Z. Ou, H. Zhuang, J.K. Nurminen, A. Ylä-Jääski, P. Hui, Exploiting hardware heterogeneity within the same instance type of amazon EC2, in: 4th USENIX Workshop on Hot Topics in Cloud Computing (HotCloud), 2012.

[90] C. Reiss, A. Tumanov, G.R. Ganger, R.H. Katz, M.A. Kozuch, Heterogeneity and dynamicity of clouds at scale: Google trace analysis, in: Proceedings of the Third ACM Symposium on Cloud Computing, ACM, 2012 (p. 7).

[91] K.K. Gopinathan, R.P. Pushpakath, S.K. Madakkara, Quantitative assessment of applications for cloud bursting, in: Advances in Computing, Communications and Informatics (ICACCI), International Conference, IEEE, 2015, pp. 1131–1136.

About the authors

Muhammad Ziad Nayyer is serving as Assistant Professor in Department of Computer Science GIFT University, Gujranwala, Pakistan. He is a full time PhD student at Department of Computer Science, COMSATS University Islamabad, Lahore Campus working under the supervision of Prof. Syed Asad Hussain. He is an active member of Communication Networks Research Center (CNRC). He received his MS degree in Computer Science from Government College University (GCU), Lahore, Pakistan in 2011. His area of intersets include Cloud Computing, VM Migration, Mobile Cloud Computing, Cloud Federation, Mobile Edge Computing, Fog Computing, and Cloudlet Computing.

Imran Raza is working as an Assistant Professor in the Department of Computer Science, COMSATS University Islamabad, Lahore Campus, since 2003. He is an active member of Communication Networks Research Center (CNRC). He holds BS (CS) and MPhil degrees from Pakistan. He has been associated with TU Ilmenau, Germany as a researcher. His areas of interests include SDN, NFV, Wireless Sensor Networks, MANETS, QoS issue in Networks, and Routing protocols.

Syed Asad Hussain is currently leading communications and networks research at COMSATS University Islamabad, Lahore campus Pakistan. He obtained his master's degree from Cardiff University, UK. He was funded for his PhD by Nortel Networks, UK at Queen's University Belfast, UK. He was awarded prestigious endeavour research fellowship for his postdoctorate at the University of Sydney Australia in 2010, where he conducted research on VANETs. He has taught at Queen's University Belfast, UK; Lahore University of Management Sciences (LUMS); and University of the Punjab. He is supervising PhD students at CUI and split-site PhD students at Lancaster University, UK in the fields of cloud computing and cybersecurity. Professor Hussain is serving as dean of faculty of information sciences and technology since 2015 and has served in the capacity of head of Computer Science Department COMSATS University Islamabad, Lahore campus from August 2008 to August 2017. He regularly reviews IEEE, IET, and ACM journal papers.

CHAPTER THREE

Toward realizing self-protecting healthcare information systems: Design and security challenges

Qian Chen
Department of Electrical and Computer Engineering, The University of Texas at San Antonio, San Antonio, TX, United States

Contents

Abstract

This book chapter reviews the history of Healthcare Information Systems (HISs), discusses recent cyber security threats affecting HISs, and then introduces the autonomic computing concept and applies the concept to design self-protecting HISs (SPHISs) that can defend themselves against cyber intrusions with little or no human intervention. To realize such SPHISs, we first study security vulnerabilities of the HIS network, communication links and protocols. Based on these vulnerabilities, the component design of a SPHIS is presented. We propose that a SPHIS should contain monitoring systems, early estimation modules, intrusion detection, network forensics analysis devices and intrusion response systems. Finally, existing self-protecting approaches for HIS, enterprise

Advances in Computers, Volume 114
ISSN 0065-2458
https://doi.org/10.1016/bs.adcom.2019.02.003

systems and industrial control systems are demonstrated in detail. This chapter provides an innovated design of an autonomic security management system that could reduce IT professional's management burdens while enhancing the security posture of their HISs.

Abbreviations

AC	autonomic computing
ADT	admission, discharge, and transfer
ARIMA	autoregressive integrated moving average
ASMF	autonomic security management framework
BCH	Boston Children's Hospital
CDA	Clinical Document Architecture
CIS(s)	Clinical Information System(s)
CPR	computer-based patient record
DDoS	distributed denial of service
DHMG	DeKalb Health Medical Group
DICOM	digital imaging and communications in medicine
DNS	domain name system
ED	emergency department
EHR	electronic health record
FIS(s)	financial information system(s)
HIS(s)	healthcare information system(s)
HIT	health information technology
HITRUST	Health Information Trust Alliance
HL7	Health Level Seven
HTTP	hypertext transfer protocol
ICMP	Internet control message protocol
IDS	intrusion detection system
IMD(s)	implantable medical device(s)
IoT	Internet of things
IoMT	Internet of medical things
IPsec	Internet protocol security
IRS	intrusion response system
LOINC	logical observation identifiers names and codes
MDP	Markov decision process
MSG	message type
MSH	message header
NCPDP	National Council for Prescription Drug Programs
NFA	network forensics analysis
NIH	National Institute of Health
NIST	National Institute of Standards and Technology
NM	Numeric
OWASP	The Open Web Application Security Project
PHI	personal health information
REDR	risk assessment, intrusion estimation, intrusion detection and intrusion response

RIS	radiology information system
RM1	Room-1
SCADA	supervisory control and data acquisition
SPHIS(s)	Self-Protecting Healthcare Information System(s)
SQL	structured query language
SSL	secure sockets layer
SSU HC	Savannah State University Health Center
SSU OE	Savannah State University Open Enrollment System
ST	string type
TCP/IP	transmission control protocol/Internet protocol
TLS	transport layer security
UDP	user datagram protocol
V2	Version 2.x
WCIS	web classifying immune system

1. Introduction

"Medical informatics is the application of computer technology to all fields of medicine-medical care, medical teaching and medical research [1]." The history of medical informatics or Healthcare Information Systems (HISs) in the United States can be traced back to the 1950s with the breakthrough of information technology and the rise of computers. The primary stage of the development of HISs was from the 1950s to 1980s. In this period, scientists developed computer software to relieve human burdens from trivial calculation and daily diagnosis.

In the 1960s billing was the center of the HISs. In the same decade, the National Institute of Health in the USA (NIH) began to support medical informatics projects, which promoted the development of HIS. As a result, administrative and financial information systems (FISs) were developed in large hospitals and academic medical centers to support the management functions and general operations of the healthcare organizations [2]. In the 1970s Clinical Information Systems (CIS) containing patient clinical and health-related information was established. Meanwhile automatized data processing and computer-assistant medical decision-making were growing rapidly. Shared systems were still used in this period, and data processing was primarily centralized on mainframe computers.

The second stage of the modern HIS was from 1980 to 2000 with the introduction of artificial intelligence and the development of the Internet and microcomputers. In this period, CIS was continuously expanded world wide, from large hospitals to small ones. FISs and CIS were finally integrated, and distributed data processing became feasible. In 1991 the Institute of Medicine began recommending healthcare organizations to implement Computer-based Patient Record (CPR) for reducing national healthcare costs while enhancing the care of patients. By the end of 2000, the adoption rate of CPR was much lower than the expectation. Only 17% of 329 family practice residency programs were using a CPR at that time [3].

The third stage of HIS was from 2000 to present when electronic health record (EHR) and e-Health systems were widely adopted by healthcare organizations. According to a recent report from the Office of the National Coordinator for health information technology, by the end of 2015, 96% of the US hospitals possessed a certified EHR technology that met the technological capability, functionality, and security requirements required by the Department of Health and Human Services [4]. From the report, one can observe that the number of hospitals implementing at least one basic EHR significantly increased since 2008, from 9.4% to 83.8%. This adoption rate includes small hospitals in the rural areas. Note that more than 50% of the increment occurred in the 2011–2015 period.

Almost in the same period, the Internet of things (IoT) applications changed the traditional healthcare industry. Today patients can remotely access their medical data by the support of cloud computing technology. The development of mobile devices and healthcare wearables allows both physicians and patients to monitor patient's health in real time for improved health outcomes. Additionally, the real-time patient monitoring and advanced homecare medical devices reduce the cost of unnecessary physician visits and hospital readmissions [5].

Although the idea of HIS was brought forward in the 1950s, the general utilization of information technology in healthcare sectors has not seen a marked increase until recent 5 years. As soon as we began to enjoy the benefits of the HIS and the Internet of medical things (IoMT) for better quality of healthcare with a low maintenance cost, the HIS has quickly become the top target of cyber criminals. According to the 2016 IBM X-Force survey of assessing and examining the goings-on in the world of cyber security and cyber threats, healthcare was the most frequently attacked industry [6]. Over 112 million health records were compromised in 2015 [7]. Some of highly publicized HIS cyber security incidents include:

1. **Phishing**: intruders masquerade themselves as trustworthy entities to obtain sensitive information by deceiving victims who are inside of

healthcare IT networks. The phishing attack is one of social engineering attacks. As an example, victims receive email messages appearing to come from legitimate enterprises but actually from attackers. Victims follow attacker's commands such as downloading malware or granting privilege to the intruders, and finally HISs are compromised and scammers gain access to the database systems and monitor specific actions [8]. Although users are very familiar with the phishing attack, it is still successful and destructive. It has already become a great risk to the healthcare industry [9]. This is because phishing attacks are cheap, easy, and flexible to launch. Moreover non-technical people such as healthcare practitioners and senior patients who are not well trained have become a pool of potential victims.

Recently the number of real-day phishing attacks that successfully compromise HISs have increased significantly. The largest healthcare cyber attack of 2015, Anthem Breach, which affected 78.8 million people, began with phishing emails sent to the employees working at Anthem INC. [10]. Phishing attacks also affected 220,000 patient information records of DeKalb Health Medical Group (DHMG), which was the eighth largest cyber attack in healthcare of 2015 [9, 11]. The ransomware attacks introduced later are one of the consequences of the phishing attacks, through which intruders gained access of the DHMG HIS, and resided in the system, monitored and obtained the organization's activities from November 2013 to January 2015. Moreover, intruders set up a fraudulent donation webpage, which was similar to the DHMG's charity donation page and sent phishing emails to ransomware attack victims for financial rewarding [12].

In recent 2 years several severe CEO phishing attacks compromised HISs. For instance, 11,000 W-2s information from healthcare workers of Pennsylvania Main Line Health was compromised via the attack in February 2016 [13]. In the same month, a similar cyber attack compromised St. Joseph's Healthcare System and 5000 employee's earning data was stolen [14]. Intruders generally masqueraded themselves as CEOs of their target organizations and sent spoof emails to the employees from a seemingly legitimate source, and eventually gained the access to patient and employee's information from the compromised HISs.

2. Malware: Computer malware is four times more likely to be seen in healthcare organizations than other industries [15]. Following the initial attack vector such as social engineering attacks, healthcare employees may download and spread malware into their HISs. According to a

study in 2015 conducted by the Health Information Trust Alliance (HITRUST), 52% of 30 mid-sized US hospitals were infected with malware, and the most common type was ransomware [16].

Ransomware attacks can encrypt all files of the compromised information systems to deny legitimate access to the data. One of the most influenced HIS ransomware attacks happened in February 2016. The attackers encrypted Hollywood Presbyterian Medical Center hospital's data and seized control of its computer systems. The ransomware attacks significantly disturbed the hospital's normal operations and paralyzed its EHR system. The hospital ultimately paid $17,000 to the hacker for restoring the functions of their computer systems [17]. Following the first ransomware attack, five similar attacks were found in the next several months. The same attack vector was used to gain accesses to information systems of hospitals and healthcare organizations in CA, KY, and Washington DC.

Incidents caused by ransomware attacks may even lead to sensitive personal and treatment data breaches. On August 24 and 25, 2016, Man Alive Inc. Lane Treatment Center was attacked by the ransomware attack that gained unauthorized access to their EHR systems, stole patients' mental health and substance abuse records, and finally sold the records on a dark web [18]. This attack highlights new challenges that the healthcare organizations can face in securing EHR containing extra-sensitive patient records.

3. SQL Injection: Attackers exploit vulnerabilities of the traditional SQL database applications used in HISs to bypass the authentication process. Intruders therefore access unauthorized Personal Health Information (PHI) that lead to sensitive data breaches. In May 2016, anonymous hackers attacked 33 Turkish hospitals databases via SQL Injection attacks and leaked more than 10 millions Turkish medical healthcare records [19, 20]. It is very likely that the same attack vector was adopted by another team of hackers who compromised personally identifiable information of 1400 employees working at York Hospital in Maine in early 2016 [21, 22].

4. Distributed Denial of Service (DDoS): This is the most common hackvisit attack overloading healthcare servers with adverse traffic to shut down EHR and email systems; thus preventing legitimate requests accessing HIS resources (e.g., critical patient information) [23, 24].

Anonymous attackers have conducted various DDoS attacks to enterprise and healthcare information systems against politicians, companies, and governments [23]. Boston Children's Hospital (BCH) was the first healthcare organization targeted by a DDoS attack in 2014 "in response to the diagnosis and treatment of a 15-year-old girl who had been removed from her parent's care by the Commonwealth of Massachusetts [23]." The DDoS attacks began with a threat and has involved three major strikes for several months [25]. The hospital immediately reacted to and mitigated the attacks with the help of third-party security companies. Even though the DDoS attacks did not significantly damage BCH HISs, their experience shows that every healthcare entity is the potential target and victim of DDoS attacks.

Healthcare and public health is one of 16 critical infrastructure sectors [26]. Unlike other computing systems, HISs are lagging in terms of proper cyber security defense and investment [27]. Medical information, however, is valuable for hackers and this makes HISs popular targets for cyber attacks. One recent survey shows that 340% more information security incidents occurred in the healthcare sector than other industries [28]. According to 2015 KPMG Healthcare Cyber Security Survey [29], only 53% of healthcare providers and 66% of health insurers out of 223 US-based healthcare organizations said that they are prepared to defend against an attack. 81% of healthcare organizations have been compromised by cyber attacks in the past 2 years.

Current healthcare organizations continuously rely on security solutions such as antivirus, firewalls and data encryption to secure their IT environment. However, HIS security is still largely at risk from sophisticated Brute Force, Phishing, Malware, SQL Injection, and DDoS attacks [30]. The biggest barriers of healthcare organizations to mitigate cyber security events are lack of financial resources and appropriate cyber security professionals. The large amount of emerging threats and the complex nature of current network infrastructure made the HIS protection even harder [30]. Therefore, there is a great urgency to develop more advanced and cost-efficient security solutions to reduce the human burden of managing and mitigating HIS threats and vulnerabilities.

Autonomic Computing (AC) technology was inspired by the autonomic nervous system that regulates and maintains homeostasis from the unconscious efforts of the brain. In an analogous way, AC technology allows computing systems and applications to manage themselves with minimum human intervention. The self-protection capacity has been applied to securing enterprise computing systems and Supervisory Control and Data Acquisition (SCADA) systems from both internal and external known or zero-day attacks with little or no human intervention [31–36]. A similar approach can be developed to supplement the system administrator's responsibility to anticipate and defend the HIS as a whole from malicious activities.

2. Autonomic computing and self-protecting HISs

The self-protection aspect of autonomic computing can be used to secure and safeguard HIS communications, properly grant authorization privileges, and evaluate performance of security mechanisms toward achieving greater resiliency. By continuously monitoring measures of performance and security, autonomic systems can ascertain differences associated with normal vs abnormal system behavior. With the usage of mathematical techniques for intrusion detection and response evaluation algorithms, the autonomic approach can efficiently and automatically respond to a wide range of cyber attacks. Since human capacities cannot possibly protect large-scale complex HISs from increasingly sophisticated cyber adversaries combined with high volumes of network traffic, the application of autonomic computing can supplement or replace human intervention needed for the management of cyber security posture and situational awareness.

Kephart [37] and Parashar and Hariri [38] summarize that an autonomic computing system must possess eight key properties:

1. Self-awareness: the computing system needs to "know itself" and is aware of its state and behaviors.
2. Self-configuration: the computing system should configure and reconfigure itself under various circumstances.
3. Self-optimization: the computing system should continuously optimize itself to enhance the operation.
4. Self-healing: the computing system should recover itself from failures.
5. Self-protection: the computing system should proactively detect and protect itself from cyber attacks.
6. Context awareness: the computing system should know its environment and act accordingly.
7. Open environment: the computing system should function within an open environment.
8. Estimated resource allocation: the computing system should optimize and anticipate its needed resources.

Therefore, to establish an autonomic computing system, one must employ control theory that provides finite methodologies for diagnosing and repairing system problems. This can be achieved with the utilization of feedback loops,

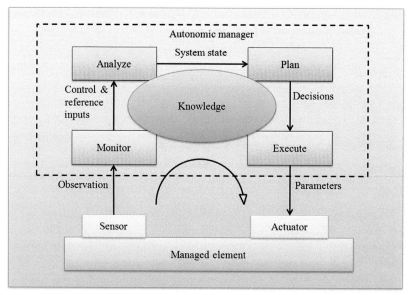

Fig. 1 Autonomic computing model.

autonomic managers, and means to communicate with surrounding environment. After analyzing this information to determine the desired (baseline) values of the observables (reference inputs), autonomic managers construct and execute plans to regulate systems based on current system states [39, 40]. Fig. 1 [39] illustrates the autonomic computing model, which is the combination of autonomic elements and feedback control theory.

In large-scale distributed environments, autonomic computing is a promising technique to enable systems to self-manage without requiring human commands. It not only successfully configures, optimizes, and heals managed elements, but also protects computer systems from cyber attacks. In the following sections, essential components, theories, and techniques employed to realize self-protecting HISs are presented. First, the control theory and its application to self-protecting systems are discussed. After that the functionality of self-protecting HISs is introduced in detail. Finally, state-of-the-art self-protecting HISs are surveyed.

2.1 The employment of control theory to self-protection

Self-protecting systems proactively anticipate cyber assaults relying on early reports generated by sensors. These systems also take steps to eliminate or mitigate cyber attacks with little or no human intervention. They adopt

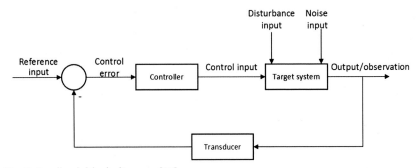

Fig. 2 Feedback block diagram [40].

the feedback control theory to monitor managed elements, analyze system performance, and select and execute appropriate plans continuously to enhance security, reduce the operator load from system management, and consequently reduce human error and cut operational costs.

The block diagram of a self-protecting system adopting the feedback control loop shown in Fig. 2 is taken from a paper by Diao et al. [40]. Similar to Diao's research, self-protecting HISs use feedback control theory to regulate user connections in the context of a constantly changing operating environment. The same type of feedback control is used to describe how self-protecting functionality is realized.

The essential elements in Fig. 2 for implementing a self-protecting HIS are described as follows:

- Target System: This refers to the victim system that is attacked by adversaries. For example, the victim can be a multitiered web application environment of the HIS, which contains web servers, application servers, and database systems; or possibly only a single database server providing patient information.
- Reference Input: Reference inputs are the desired values from target system outputs. In reference to security aspects, the desired system outputs are: (1) providing "normal" services to legitimate clients and (2) eliminating and protecting systems from all cyber assaults.
- Control Input: Control inputs are effective countermeasures or responses that are able to eliminate or mitigate cyber attacks thereby improving the security of HISs within their specific operating environments.
- Control Error: Control errors result from large differences between the *Reference Input* baseline and current observations. The differences could be Euclidean distances between measured system states and the "normal region" of a secured system. The differences also provide the basis for

determining the likelihood that the system performance is within the "normal" operating range.

- Controller: The controller evaluates recommended responses or policies, and initiates the most appropriate responses needed to recover the compromised HISs considering control environment inputs.
- Disturbance Input: Any cyber attacks undermining confidentiality, integrity, and availability of HISs are represented by disturbance inputs. Disturbance inputs may include malicious requests, unauthorized accesses, spoofing, DDoS attacks, and eavesdropping (i.e., man-in-the-middle).
- Noise Input: This refers to missing data or noisy data collected by sensors.
- Transducer: It transduces system security states in real time to enable the security states of the *Target System* to be compared with the *Reference Inputs* (i.e., "normal region") to ascertain the Control Errors.
- Output/Observation: The output of the controller is the observation of the *Target System* security states, which is composed of real-time values such as system, network, and security features of the servers or power consumption of medical devices.

2.2 Self-protection model and functionality

To better understand how a self-protecting HIS is architected, we developed a comprehensive autonomic security management framework that contains Risk assessment, intrusion Estimation, intrusion Detection and intrusion Response (REDR) modules. Fig. 3 shows the REDR framework that gives a model for adopting control theory to assess vulnerabilities and to baseline the normal region of the HIS behavior for a specific operational environment. The self-protecting HIS, as the first-line of defense, anticipates

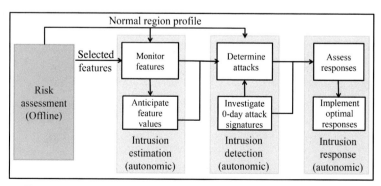

Fig. 3 REDR an autonomic security management model.

upcoming anomalies, sends early warnings to the controller, and protects the system by signaling the controller to initiate an active response to selected threats. Sophisticated attacks that evade the first-line of defense due to incorrect estimations of future system security states may be thwarted by an intrusion detection system (IDS). Live network forensics analysis is used to learn the causes and impacts of unknown attacks. Novel signatures [33, 41, 42] are utilized to update detection algorithms and the active response library. As a result, similar attacks can be removed in the future by the first-line of defense. The functionality of each module is discussed as follows:

- Risk assessment: The first step to establishing a self-protecting HIS is to assess its security risks offline. Risk assessment helps an organization to determine the impact and likelihood that a given threat will and can successfully exploit a particular vulnerability. HITRUST Risk Analysis Guide introduces a risk management framework in four steps [43]. In our REDR model, the risk assessment module is divided into five steps (as shown in Fig. 4). We added the component of 'characterize HIS' before the four steps guided by HITRUST. This component analyzes and characterizes the HIS performance with the help of expert knowledge offline, followed by the establishment of the "normal region," which represents secure system performance using normal datasets of relevant features. These features can be parameters such as CPU time, memory utilization, I/O read/write rate, packet received/sent rate, TCP connection performance, communication protocol message header and payload that impact HIS' system, network, and security resources.

  System administrators can identify risks to their HISs in step two. Outputs from this module indicate the degree to which the HIS is vulnerable. This step also must determine protection requirements to minimize high-risk areas of HISs. The third and fourth steps are

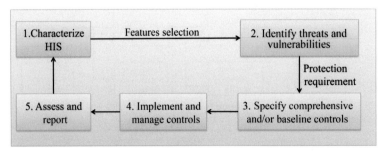

Fig. 4 Risk assessment steps and modules.

to select, implement and manage comprehensive and/or baseline security controls based on the risk assessment result, which in turn prioritizes the active response to abnormal unwanted or unauthorized activities. Examples of these controls include firewalls, antivirus software, access control lists, security policies, and honeypots.

The last step is to assess the controls selected and implemented in steps three and four, and report the efficiency of the selected controls for mitigating risks to business associates and system administrators. If these controls cannot prevent HISs against known risks, more efficient controls must be reselected and implemented.

Features monitored by the intrusion estimation module are the same parameters used to determine the "normal region." Real-time observations are collected by the system monitors.

- Intrusion estimation: System sensors continuously monitor selected features. Estimation of the system security state uses historical observation to project future feature values using statistical methods combined with system models. Upcoming attacks are therefore anticipated by comparing the "normal region" with the estimated outputs of the system model using forecasting methods (e.g., time series forecasting methods such as Kalman filter or the autoregressive integrated moving average (ARIMA) method [33, 44]). Early warnings, therefore, are sent to the controller, and appropriate control mechanisms are executed to neutralize attacks (on-the-fly).

- Intrusion detection: The IDS is a data mining tool that enables real-time event analysis. This module provides accountability for intrusive activities thereby detecting anomalous system behavior. To this end, three types of IDSs (i.e., host-based, network-based, and hybrid IDSs) must be developed using signature and anomaly techniques. The anomaly detection technique comparing real-time system performance with the normal system model detects known and unknown attacks. The signature detection technique identifies known attacks relying on matching observations to known misuse patterns. To identify zero-day attacks (i.e., attacks that exploit previously unknown vulnerabilities using perhaps unknown tactics and methods) the signature database must be upgraded frequently. Using appropriate detection techniques can identify both known and unknown attacks with low false alarm rates. In Fig. 3, the intrusion detection module has the ability to analyze evolving zero-day attacks. This functionality is realized using live forensics analysis tools to learn unknown attack signatures. These signatures are adopted by online detection algorithms and defense mechanisms.

As a result, the previously unknown attacks can be autonomously incorporated into the corpus of known attacks used subsequently by the first-line of defense.

• Intrusion response: The utilization of intrusion response systems (IRSs) enables the self-protecting HIS to mitigate attack impacts and regulate system behavior back to normal. IRSs can be divided to static and dynamic types. A dynamic IRS is commonly used to enable an ASMF system. This is because it allows the execution of lower ranking responses for the protection of the compromised system if the so-called optimal responses are not sufficient to mitigate attack impacts. If lower ranking responses are implemented and successfully regulate system performance, the dynamic IRS will escalate rankings of these responses by modifying the initial values of the assessed criteria. Examples of relevant criteria include response effects (e.g., how fast can responses regulate system behavior back to normal), operational costs, and impact on system's services.

Feedback control theory is the primary means used to realize self-protecting HISs. Therefore, after the self-protecting system realizes all of the steps mentioned above, sensors will still continuously monitor selected features to identify whether system performance does actually return to normal running behavior. When abnormal behavior reoccurs, the system will repeat the self-protection processes mentioned here until the system reaches the baseline normal operational range of states.

3. Healthcare information system

Before we introduce how to realize a self-protecting HIS, let us first understand the computer network, components and communication standards of a typical Healthcare Information System (HIS). The HIS normally contains FISs and CISs. Similar to an enterprise system, a HIS adopts a three-tier client–server architecture that contains a web tier, an application tier, and a data tier for retrieving and storing healthcare related information. As shown in Fig. 5, the web tier, also known as the presentation tier, gives patients and health providers access to a server-based application visible from the graphical user interface via desktops, tablets or mobile phones. The application (logic) tier contains mobile and web applications, which manages business logic, processes commands, and transforms data. The data tier collects operational and business data from CISs and FISs and stores the data in databases or repositories.

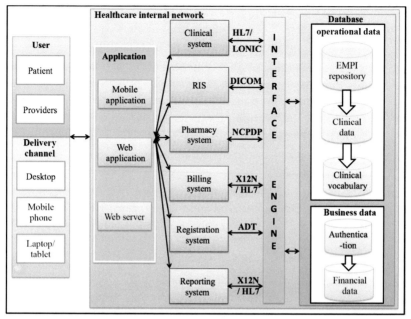

Fig. 5 Healthcare information system and communication protocols [45].

3.1 Communication standards

Communication standards for clinical, image, prescription and administrative data transmission are different. Fig. 5 presents a clinical system, radiology information system (RIS), pharmacy system, billing system, registration system, and reporting system. Relevant data of these FIS and CIS are collected and stored in the clinical and financial databases. To communicate with the data tier, each information system must follow individual protocols. HL7, DICOM, NCPDP, X12N, ADT, and LONIC are communication standards widely used for healthcare CIS and FIS. These standards are introduced as follows:

- Health Level Seven (HL7) is an interoperable standard allowing disparate healthcare applications to exchange, integrate, share and retrieve key sets of clinical and administrative data [46]. Therefore, data transmission between the clinical system, billing system, reporting system, and the databases should follow the HL7 standard. More details of the HL7 messaging standard is presented in Section 4.
- Digital imaging and communications in medicine (DICOM) regulates the standards for exchanging digital radiology and cardiology images between RIS and other CIS [47].

- The National Council for Prescription Drug Programs (NCPDP) creates standards for electronic communication of prescription, billing, and other pharmacy material. The NCPDP SCRIPT Standard "supports messages regarding new prescriptions, prescription changes, refill requests, prescription fill status notification, prescription cancellation, medication history, and transactions for long-term care environment." The NCPDP Telecommunication Standard "supports the electronic communication of claims and other transactions between pharmacy providers, insurance carriers, third party administrators, and other responsible parties [48]."
- X12N is the standard that enables the electronic exchange of healthcare insurance, claims processing data and billing information. Similar to the HL7 standard, X12N can be used for retrieving data from billing systems and reporting systems.
- Admission, Discharge, and Transfer (ADT) messages carry patient information such as patient ID, medical record number, age, name and contact information. ADT messages can also provide "trigger events such as patient admit, discharge, transfer, registration, etc [49]." Therefore, in Fig. 5, the registration system transfers patient information via ADT messages.
- Logical observation identifiers names and codes (LOINC) is the standard for identifying medical laboratory observations. Clinical and laboratory LOINC standards enable the electronic exchange and gathering of clinical results such as laboratory tests and research. After that, the clinical results are recorded into clinical reports and documents.

4. HL7 messaging standard

This section introduces one of the most widely used HIS communication standards, Health Level Seven (HL7), and its security risks. HL7 was developed by a nonprofit organization providing a comprehensive framework and developing standards for the communication of electronic health information [50].

HL7 was founded in 1987. Currently, it has more than 2500 members representing 90% of HIS vendors from 55 countries. The recent report [50] shows that 95% of US healthcare organizations are using the HL7 Version 2.x (V2) messaging standard for exchanging health information in the clinical domain. HL7 V2 is also considered as one of the most widely implemented standards for healthcare in the world. From HL7 V2.5 to the latest HL7 V2.9

released in 2016, these HL7 messaging standards are foundational and mandated by the US Meaningful Use legislation, which means that adopting HL7 standards can "best inform clinical decision at the point of care [51]."

The 2016 Interoperability Standards Advisory (2016 Advisory) recently published a report of the best available standards and implementation specifications [52] based on two rounds of public comment and recommendations from the health information technology (HIT) Standards Committee. The report evaluates HIT standards, and categorized the evaluations in three sections, and they are: best available vocabulary/code set/terminology standards and implementation specifications; best available content/structure standards and implementation specifications and best available standards implementation specifications for services.

HL7 V2.5.1 was named the best available standard for admission, discharge, and transfer; interoperability need, receiving electronic laboratory test results, ordering labs for a patient, etc. In addition, HL7 V2.5.1 is highly adopted in health care within the United States. Therefore, the security of communication under the HL7 V2.5.1 has become a significant challenge. In the following sections, we first introduce the structure of HL7 V2.5.1 data frames in Section 4.1 to have a better understanding of the security risks of the HL7 V2.5.1 message standard, and then we create a message sent from a client (such as from a doctor's office) to the servers to illustrate the meaning of the MSH's (message header) fields. We also discuss the security challenges in HIS with the HL7 MSH message as an example (shown in Section 4.2).

4.1 HL7 V2.5.1 data frame

HL7 V2 message standards contain a segment and some fields. The content of fields is primitive data types such as Numeric (NM) for sequence numbers and String Type (ST) for field separators. The second type of the field is the composite data type, which contains one or more components. For example, message type (MSG) is a composite data type containing three ID types such as message code, trigger event, and message structure IDs. The fields of the HL7 message can be repeated.

An example of the HL7 message is shown as follows:

HL7 V2.5.1 contains 144 segments covering all clinical and financial activities of the patients, doctors (hospitals), pharmacies, and insurance companies [53]. The segment of the example in Fig. 6 is "Message Header" (MSH). The MSH segment is shown in every HL7 message type and defines source, purpose, destination and specifics of the syntax of the messages.

MSH|^~\&|SSU HC|LAB-1|SSU OE|RM1|201605021030||VXU^V04^1|SSUCTL-1|P|2.5.1
|M2 |M3 |M4 |M5 |M6 |M7 ||M9 |M10 |M11|M12

Fig. 6 An example of a HL7 V2.5.1 MSH message and its fields.

Besides the segment, the HL7 V2.5.1 MSH messages contain up to 21 fields counting from MSH.1 to MSH.21. In Fig. 6, we wrote MSH.2 as M2 for short to save space. Seven of these fields are required, and they are described as follows:

1. MSH.1 defines the field separator (|). In Fig. 6, we did not name M1 since MSH.1 is the field separator "|", and it is also used as the separator to separate MSH.1 and MSH.2.
2. MSH.2 defines other separator characters for the message in the order of: a component separator (∧), a field repeat separator(~), an escape character separator (\) and a subcomponent separator(&).
3. MSH.7 defines the date or time of the message. In Fig. 6, the time for sending the MSH message is May 2, 2016 at 10:30 am.
4. MSH.9 defines the type of message and trigger events. In the sending application, "VXU" is the message code that shows that the message contains unsolicited vaccination record update information, and the trigger event number, "V04," matches the message code. "1" is the message structure. These three components are separated by component separators (∧).
5. MSH.10 defines the message control ID, which is a number or unique identifier for the message. As shown in Fig. 6, we created 'SSUCTL-1' as the control ID, which is a string type.
6. MSH.11 defines the processing ID. In Fig. 6, the processing ID is "P".
7. MSH.12 defines the HL7 version ID. Version IDs are used for message transmission and help the server side to interpret the messages with an appropriate HL7 version. Since we selected the HL7 V2.5.1 MSH message as an example, the version ID shown in Fig. 6 is 2.5.1.

The example MSH message shown in Fig. 6 also contains another four optional fields from MSH.3 to MSH.6. The description of these fields are as follows:

- MSH.3 defines the sending application. Fig. 6 shows that this message sends from Savannah State University Health Center (SSU HC).
- MSH.4 defines the sending facility. Fig. 6 shows that the sending facility is Lab-1, which can be an immunization and testing lab.

- MSH.5 defines the receiving application. In Fig. 6, the receiving application is Savannah State University Open Enrollment System (SSU OE).
- MSH.6 defines the receiving facility. Fig. 6 shows that the SSU OE system is located in Room-1 (RM1).

More details of another ten optional fields of MSH messages and other 143 segments can be found in this website [53].

4.2 Security challenges in healthcare information systems

Most healthcare organizations rely highly on commercial security products. However, due to practical limitations and unique drawbacks of these security solutions, HISs are far from attack-proof. Scanning attacks, injection attacks, broken authentication and session attacks and DDoS attacks are most dangerous HIS threats. In this section, we first introduce the features and adverse impacts of these attacks, and then we use the HL7 MSH message shown in Fig. 6 to illustrate how these attacks will be simulated, detected, and mitigated.

1. Scanning attacks: adversaries scan devices in HIS to gather network information of these devices before launching sophisticated attacks to undermine the HIS security. Commonly used scanning techniques to gather computer network information include IP address scanning, port scanning, and version scanning.

   Besides the normal IP address and port scanning techniques (e.g., address resolution protocol and TCP SYN scans) to gather the server IP addresses and the opening ports for the HL7 message transmission, adversaries can also carry out segment scanning attacks to learn message, personal identifiers, order numbers, or patient visit information of specific HL7 messages. Malicious hackers can also obtain the HL7 versions by eavesdropping the version ID field of the MSH messages.

   To protect the HIS against scanning attacks, it is better to close ports that are not commonly used. In addition, relative intrusion detection and protection rules must be updated to drop malicious requests sent to the ports that HL7 servers are listening on.

2. Spoofing attacks: Spoofing attacks happen when adversaries pretend to be legitimate users. Masquerading and impersonation are two types of spoofing attacks. Masquerading is a passive attack. Attackers first exfiltrate legitimate account credentials and then log in to the HIS as legitimate users. Impersonation is also known as replay attacks. This spoofing attack is more active than masquerading. Attackers

capture authentication traffic and replay that traffic to gain access to the HIS. Once adversaries control the compromised HIS, they can exfiltrate confidential data, exhaust system resources, or spread malware to compromise other networks.

To detect and protect the HIS against spoofing attacks, server sides must frequently change the authentication credentials and enable duplicate detection techniques.

3. Injection attacks: In the data tier, adversaries can exploit the vulnerabilities of structured query language (SQL), JavaScript, and computer programs to trick the interpreter by inserting untrusted data. As a consequence, adversaries may obtain access to the database, attack web users, and propagate computer worms. On the other hand, in the HIS network, adversaries may inject malicious segments, commands, or responses to reduce the security of HIS.

To this end, the HIS messages must be encrypted, and the senders must pass the authentication. The HIS can also avoid SQL queries from users or switch to nonSQL databases.

4. Broken authentication and session management: Attackers exploit vulnerabilities in authentication mechanisms to assume legitimate users' identities. A successful attack may compromise all accounts. As a result, adversaries can do whatever legitimate users could do. The brute force attack is an example. It takes advantage of weak passwords and small encryption keys. Such attacks send guessed values of account usernames, passwords to the server. Attackers may repeat sending their guesses until they successfully access system resources from compromised accounts.

The EHR servers must strictly limit the account access attempts to avoid brute force attacks. Meanwhile, the IP addresses of the malicious users must be added to the detection rules for filtering malicious packets in the future.

5. Distributed denial of service attacks: DDoS attacks are intended to make HIS resources unavailable using various methods and tactics to exhaust system and network resources. One example is flooding-based DDoS attacks that send a tremendous number of packets to swap web servers. In this way, bogus requests block legitimate requests by overwhelming system's resources (e.g., CPU, memory, bandwidth).

The best way to protect the HIS from DDoS attacks is to hide the systems (e.g., IP addresses) from users. The flooding messages can be redirected to honeypots. The information of attackers such as IP addresses, protocol types, and source port information must be updated for the detection and protection rules.

5. State-of-the-art research for enhancing HIS security

Sections 3 and 4 introduce HIS networks, communication standards and the potential security threats. In this section we present state-of-the art research on enhancing HIS security. The HIS can be categorized as an enterprise system and hence, the self-protecting feature in HISs is still in the initial stage. Therefore, besides surveying research on enhancing HIS security posture, we also reviewed existing technology to protect information systems in relevant industries (i.e., distributed enterprise systems and supervisory control and data acquisition (SCADA) systems).

Self-protecting systems know their environment and proactively enhance network and information security. As described in Fig. 3, before a self-protecting system can be useful, the system boundary (e.g., routers, firewalls, web servers, application servers, databases) must first be determined so that those security controls can be baselined during the risk assessment phase. Abnormal system behavior is then a quantifiable deviation from the baseline. Anticipated abnormal system behavior is the early stage of a departure from the baseline. Self-protecting HISs (SPHISs) compare each new request to the anticipated behavior, followed by the execution of an appropriate control. One example of self-protection in a distributed system is introduced by Claudel et al. [54]. Data from the sensors and actuators of the proposed system are utilized to identify abnormal (malicious) requests and isolate compromised nodes automatically. Chen et al. [32, 33] adopted the idea of autonomic computing and created self-protecting systems that can estimate known attacks before they impact devices in the network. If an attack bypasses an intrusion detection system (IDS), a second line of defense provides patterns of known attacks to an intrusion response system (IRS). A controller in the IRS evaluates candidate protection mechanisms considering predefined criteria and then selects the optimal protection mechanisms to regulate system behavior back to normal. Unknown attack patterns were investigated by an online-learning module and were sent to the IRS similar to the known attack scenario. As a result, the self-protecting systems were validated to protect SCADA systems against both known and unknown attacks in near real time with little or no human intervention.

Compared with various self-protecting systems implemented in enterprise and industrial control systems, research on applying autonomic computing to secure HISs is not very mature. Hsieh [55] proposed a framework integrating access control policies, digital signatures and encrypted data into the Clinical Document Architecture (CDA) to realize self-protecting e-Health CDA documents. This work enhanced patient data integrity and privacy, however, the self-protecting functionality is very limited. The system cannot react to breaches or learn novel attack patterns. Another group of researchers extended the autonomic computing concept and created a self-managed healthcare Emergency Department (ED) system. This autonomic computing system reduces the dependency on human management and improves system operations and decreases service cost [56]. Lupu et al. proposed a self-managed cell as an architectural pattern for a ubiquitous computing environment to monitor patient's health in hospitals or at home [57]. However, neither of these two designs include the self-protection feature; therefore, the self-managed ED systems and the healthcare monitoring system are very likely to be attacked.

Although none of current self-protecting approaches have been fully realized as a complete security design for HISs based on the autonomic computing concept developed by Kephat et al. [39], the techniques and approaches used in other computing systems can help us realize a SPHIS. For example, we can adopt the perfect low-cost monitoring system developed by Salih and Zang [58] to contentiously collect patient health data or monitor electrocardiogram data obtained in real time using the Pandey et al.'s design [59]. We can also study from multidisciplinary approaches presented in [60] to achieve an efficient and trustworthy e-Health monitoring system.

In Sections 5.1–5.3, we review the intrusion detection, network forensics analysis and intrusion response technologies that can be applied to secure HISs and eventually realize a self-protecting HIS.

5.1 Intrusion detection systems

An intrusion detection system (IDS) is a data mining tool used to identify cyber attacks. Besides quickly identifying attacks, it has many other benefits such as enabling the collection of intrusion information, recording malicious events, generating reports, and alerting system administrators by raising an alarm.

Recent literature shows that there are several challenges to overcome for today's IDSs to achieve their goals. The first challenge is the slow speed at which IDSs detect anomalies, especially for signature-based IDSs that match observations to misuse patterns. The second challenge is that most prevalent IDSs cannot detect or adapt automatically to novel zero-day attacks. Most existing IDSs deployed in HISs use a signature-based technique. Known attacks can be easily classified based on attack patterns but not zero-days. On the other hand, by adopting the "anomaly detection technique," IDSs can detect both known and unknown attacks. However, without proper classification, deploying accurate counter measures is impossible. The third challenge of current IDS research is the high rate of false alarms. An IDS is an important component in a comprehensive security design, and many security solutions cannot be deployed without an accurate attack detection capability. Moreover, inaccurate detection will adversely affect response selection by the IRS. The obvious result will provide attackers with greater opportunity to compromise the whole system. The following paragraphs overview recent research addressing these three challenges.

The bottleneck of detection speed often plagues signature-based IDSs. This type of IDS must match traffic patterns to a list of rules. The detection speed is influenced by the length of the shortest pattern of rules and the memory access latency [61]. To accelerate the detection speed, recent research has deployed a variety of fast pattern matching algorithms such as the modified Wu–Manber algorithm [62], the set-wise Boyer–Moore algorithm [63], and the FNP algorithm introduced in [61]. Tan et al. [64] extracted partial sample strings from patterns of attacks to enhance the implementation speed.

Using modern graphics processors is another efficient method to address the first challenge. Gnort is an example that yields a detection speedup from two to ten times [65]. Researchers demonstrated that a vulnerability specification and its optimized architecture called VESPA [66] can reduce the time to detection. VESPA uses design principles like fast primitive matching, explicit state management, and parsing of relevant message parts.

The second challenge of today's IDSs is the inability to detect zero-day attacks or classify known attacks. Bolzoni et al. [67] demonstrated an IDS called Panacea, which integrates anomaly based with signature-based IDSs using machine learning methodologies to address these challenges. The adoption of support vector machine and RIPPER rule learner techniques allows Panacea to classify known and unknown attacks. Another way to enhance the detection rate is to use a behavior rule specification-based IDS for detecting unknown attacks against medical devices embedded in a medical cyber physical system [68]. Boggs et al. [69] examined user-defined content from HTTP requests to detect zero-day attacks across multiple web servers. The abnormal requests were identified by local detection sensors. This approach is very efficient at detecting previously unseen attacks with a false positive rate of only 0.03%. However, this method only examines HTTP attacks and cannot detect attacks that exploit vulnerabilities of other network protocols (e.g., HL7, TCP, ICMP, or DNS).

Amann et al. [70] presented an input framework that integrates external information into the IDS decision process. External information is the third-party intelligence that includes known botnet servers, malware registries, and blacklists. Frequently updating external intelligence does not cause a delay in detecting anomalies; however, this framework may bring additional security concerns from attackers who game the system. Artificial immune systems inspired by their biological counterpart have been applied to scan network traffic and identify zero-day attacks. Danforth [71] developed a web classifying immune system (WCIS) employing an artificial immune system, which achieved a high rate of accuracy for identifying and classifying SQL injection attacks, XSS attacks, and buffer overflow attacks.

The third challenge of IDSs is how to improve the accuracy of the detection rate (i.e., reduce false positive/negative). In a large-scale distributed system, the choice and placement of detectors influence the accuracy of the overall detection function. Modelo-Howard et al. [72] used a Bayesian network model to evaluate the effects of detector configuration and placement against the accuracy of the detection rate and the likelihood of an attack goal being achieved. Selecting the correct features to monitor is another key to improve detection rates. Chu et al. [73] built a user behavior model adopting statistical methods based on device access logs to catch anomalies and protect the networking infrastructure. The user behavior model contains several aspects that can best distinguish normal activities from attacks. These aspects contain failed login attempts, login access patterns, and user behavior. Researchers have shown that the IDS with log analysis approach effectively identified potential intrusions and misuses with an acceptable overall alarm rate. Continuous tuning of the intrusion detection model can be used to enhance the accurate detection rate, and Yu et al. [74] has presented a case in point. The paper introduces the ADAT IDS that uses a buffer for holding suspicious predictions. Only the predictions whose confidence values are higher than predefined thresholds are sent to the system operators for identifying false predictions. The feedback is then used to tune the detection model so that misclassified attacks are dropped resulting in higher detection accuracy.

From the literature review, we can summarize that an IDS that quickly detects and classifies known and unknown attacks with a high detection accuracy is one of the most essential components of a SPHIS. The output of this module provides the security state of the system. If the system is compromised, network forensics analysis must run to learn unknown attack signatures, followed by internalizing the necessary responses needed to recover system performance and defend against such attacks.

5.2 Network forensics analysis (NFA)

Network forensics aim at finding out causes and impacts of cyber attacks by capturing, recording, and analyzing of network traffic and audit files [75]. NFA helps to characterize zero-day attacks and has the ability to monitor user activities, business transactions, and system performance. As a result, attack attribution, what attack methods and tactics were used and attack duration, can be analyzed and derived using NFA.

Recent articles claim that self-protecting systems are able to mitigate attack impacts by "static" IRS. Protection mechanisms selected by this type of IRS remain the same during the entire attack period [76]. None of these self-protecting systems, however, contain NAF. Notwithstanding, a static IRS is limited in its ability to fully and efficiently eliminate these attacks. To establish a fully autonomic computing system, as described in Section 2.2, the NFA must be activated once unknown attacks have been identified by the IDS. Some recent research performs the NFA as a postprocess analysis, but live NFA is required to support the 24/7 system availability. In this way using live NFA, the investigation of unknown attack signatures will not disrupt normal client requests. Additionally, live NFA is executed immediately after unknown attacks are detected, which reduce latency and quickly closes the window of vulnerability.

Gunter [77] discussed research on balancing access controls with audit in medical record systems, and using audit logs to understand roles and workflows to develop reliable anomaly detection and safe access controls. Philli [75] illustrated a generic process model for network forensics in real-time and postattack scenarios. The first phase is preparation, which makes sure security solutions such as monitors, firewalls, and IDSs have already been deployed. The second phase is detection, which identifies cyber attacks and captures network traffic, followed by the third incident response phase. The fourth phase is collection and preservation. Traffic data are collected and preprocessed for storage on backup devices for presentation. Together these phases identify attack indicators, classify attack patterns, determine attack paths, and provide documentation offline.

The challenges of next generation state-of-the-art NFAs are described by Philli [78]. The analytical complexity of real-world large-scale HISs grows exponentially, therefore, making full NFA computationally infeasible. In some NFA scenarios accurate vulnerability states are not available by virtue of the fact that the NFA process is fundamentally postcompromise. Also, analyzing an extremely large amount of auditing data in a very short time span is computationally infeasible. Additionally, in NFA it is often very difficult to distinguish primary attack signatures from trivial signatures. Let us examine today's live NFA tools.

Investigating high-volume auditing data in real time depends primarily on the abundance of computational resources. Normally, the size of one-days worth of auditing files for large-scale systems is larger than 100 GB. Data concentrators are used, for example, by data historians within the context of the distribution systems (i.e., the smart grid). Cloud storage platforms are, used to store and process such data. Chen [79] used the cloud storage and computing platform to analyze offline phishing attacks. Phishing filter functions can effectively scale to detect phishing attacks. Research described in [80, 81] adopts the Hadoop map reduce model to analyze a high volume of log files and extracts attack signatures for intrusion detection and prevention.

Contemporary NFA approaches manually investigate attack patterns after HISs are compromised. Real-time and dynamic forensics systems, which investigate attack causes and impacts automatically, have been deployed to address the scalability issue of such manual processes. The utilization of the cloud-based approaches makes NFA more practical to deploy. Nonetheless, the current question about how to realize a dynamic and automatic NFA system is the subject of much research. For example, Wang et al. [78] developed immune agents, which automatically generate crucial evidence of unknown attacks for rapid active responses. An automatic method was proposed by Wang et al. [82] to make regular expression signatures of the HTTP attacks offline. This approach first extracts application session payloads to identify common substrings and their positional constraints. Using these extracts, attack signatures are transformed into regular expressions.

Li et al. [83] designed a network-based signature generator, namely, LESG. The generator automatically analyzes zero-day buffer overflow vulnerabilities and generates length-based signatures via a three-step algorithm. The generator selects field candidate signatures, optimizes the signature lengths, and derives the optimal signature that produces minimal false alarms. The LESG has lower computational overhead than research by Brumley et al. [84], which generates vulnerability signatures using regular expressions. False alarm rates of an IDS applying the LESG are significantly reduced and the attack detection speed is improved. The LESG is also tolerant to noisy traffic.

The utilization of NFA allows the self-protecting system to investigate and learn the causes and impacts of an unknown attack. In the end, the HISs can be protected from both known and unknown attacks autonomously (i.e., without human intervention).

5.3 Intrusion response systems

An intrusion response system (IRS) is a critical part of the self-protecting system for ensuring appropriate responses are dispatched to react to protect the HIS and recover system performance back to normal. The tight interaction between the NFA module and the IRS module makes future attacks, which have similar signatures, less likely to succeed [85]. The development of an autonomous IRS focuses on two key elements: configuring suitable responses and evaluating recommended responses dynamically.

Responses are protection mechanisms that have the ability to eliminate or mitigate cyber assaults. Protection mechanisms, which are configured and implemented to defend the system against specific cyber attacks, vary according to the types of cyber attack they target. In other words, recent research shows that using key-agreement schemes and a heart-to-heart authentication system are meaningful approaches to secure implantable medical devices (IMDs) [86, 87]. However, these responses are not efficient in removing DDoS attacks or SQL injection attacks. Five categories of cyber attacks are illustrated in Section 4.2: scanning attacks, spoofing attacks, injection attacks, broken authentication and session management, and DDoS attacks. In the subsequent paragraphs, responses that protect HIS web applications from these attacks are reviewed.

Scanning attacks are conducted by adversaries to discover vulnerable services. Port scanning, IP addresses scanning, and version scanning attacks are discussed in Section 4.2. Techniques such as TCP scanning, UDP scanning, and ICMP scanning gather the IP addresses of victim hosts looking for open ports. To protect HISs from scanning attacks, packet filters are placed in the victim system to discard scanning packets. Transparent proxies, which hide actual open ports from external users, are also somewhat effective mechanisms to defend the system from scanning attacks. A successful spoofing attack will utilize legitimate identity credentials to enable or bypass authentication mechanisms and unauthorized remote system access. Securing Internet protocol communications by authenticating and encrypting IP packets is a promising defense measure (e.g., using the Internet protocol security (IPsec)). Another example is secure socket layer and transport layer security (SSL/TLS), which uses public key infrastructure to secure transport layer transactions between clients and servers [88–90]. Mechanisms like secure remote password protocol and dynamic security skins are generally used by transaction based web sites to defend against authentication attacks [91, 92]. In addition, both access control lists, which block communications between malicious clients and servers, and packet filters, which reject spoofing attack packets are useful spoofing countermeasures.

Injection attacks insert illicit data into web applications and force victim systems to execute undesirable codes or commands and spread malware. One of the best methods to protect from injection attacks are "secure coding best practices." The Open Web Application Security Project (OWASP) lists 12 secure coding practices such as input validation, output encoding, and file management checklists [93]. Limiting web application coding privileges, reducing debugging information, and testing web applications regularly can also mitigate injection attacks [94].

Brute force attacks and man–in–the–middle attacks are broken authentication and session management attacks, which exploit flawed credential management functions. Once attackers successfully break authentication mechanisms or hijack active sessions, it may be possible for them to gain access to all accounts and do anything legitimate users could do. To protect against broken authentication and session management attacks, a simple but efficient method is to block IP addresses. Another simple and effective practice is to create and follow security guidelines for choosing strong passwords. No default usernames and/or passwords are allowed. Disabling remote logins via the secure shell and creating a large number of honeypot accounts in financial systems is a common practice used to mitigate brute force attacks [95–97].

Flooding-based DDoS attacks as discussed above (Section 1 and Section 4), aim at impeding legitimate clients from accessing host/server resources. There are numerous traditional mechanisms the host can employ to defend against DDoS attacks. Mudhakar et al. [98] used a twofold mechanism to filter application layer DoS attacks. This approach uses hidden ports to filter illegitimate packets. Then web server resources are allocated to admit clients based on priority levels assigned by the congestion control element. This protection mechanism works well for external DDoS attacks, but it is ineffective in eliminating internal DDoS attacks (i.e., compromised internal nodes initiate the flood). Performance overheads from applying this approach are very high resulting from the necessary adjustments needed to employ hidden ports and to monitor/manipulate client priority levels.

Stavrou et al. [99] introduced WebSOS, an adaptation of the Secure Overlay Services (SOS) architecture that only allows legitimate users to access web servers when DoS attacks have compromised the server. Patil and Kulkarni [100] proposed a lightweight mechanism that uses trust to differentiate between legitimate users and attackers who launch HTTP flooding attacks. However, it is unclear whether this mechanism will reduce high utilization of web server resources caused by application layer DDoS attacks.

The first key element of an IRS is selecting and configuring protection mechanisms. Any of the above mentioned mechanisms can eliminate or at least mitigate the variety of potential attack tactics. The second element is a way to evaluate these candidate response mechanisms for choosing the most appropriate mitigation/recovery response. The most effective response, however, may not protect a system from other attack types. To accurately evaluate these responses, Luo et al. [101] designed a game-based multistage intrusion defense system, which allows administrators to evaluate which type of responses are optimal with respect to different measures (i.e., selectivity, rapidity).

The optimal responses are the ones with the highest utility values based on administrator estimations of expectations and variance. Another useful evaluation method is to employ cost-sensitive modeling to select optimal responses. This method accounts for the balance between an estimated cost of potential damage vs the cost of implementing a particular set of responses [102, 103]. Chen et al. [33, 34, 44] developed a multicriteria analysis controller for the IRS to select the most appropriate responses consider their protection functionality and overheads. The researchers demonstrated that their approach could react to various cyber attacks to distributed systems and SCADA systems, and the IRS regulated system behavior back to normal in real time.

Dewri et al. [104] addressed a multiobjective optimization problem for how to secure systems with limited budgets. The researchers created a system attack tree model to evaluate measures based on total security control costs and residual damages. This approach provides concrete options to the system administrators about the quality (i.e., cost) of the different trade-off solutions. However, system administrators are not allowed to modify their decisions during run-time. Stakhanova et al. [105–107] evaluated finite responses for three criteria: response system impacts, response goodness, and response operational costs. Their response evaluation relies on expert knowledge, but the evaluation cannot distinguish which action is the best if all responses have similar descriptions.

Designing dynamic IRSs must account for installing and configuring suitable protection mechanisms as well as selecting corresponding evaluation methods. The dynamic response system is an essential module in a self-protecting system that actively reacts to (thwarts) attacks after they are detected by the IDS. This approach reduces the time gap between attack detection and response, which reduces the (time) window of opportunity for the attacker. Iannucci et al. [108] present a dynamic IRS using Markov decision process (MDP) to evaluate its long-term response policies according to a multicriteria objective function. The proposed IRS was validated to protect large systems while introducing little or no overhead on the protected hosts.

Additionally, to protect HISs against cyber attacks is not limited to the technology aspect but also should include security laws, regulations and guidelines as well as establishing a comprehensive reporting system to capture security-related failures in medical devices [109].

6. Conclusion

This book chapter first reviewed the history of healthcare information systems (HISs), introduced and analyzed recent emerging cyber breaches that affected HISs, and then proposed the autonomic computing concept and applied the concept to design self-protecting HISs (SPHISs) that can defend themselves against cyber intrusions with little or no human intervention. To realize such SPHISs, we studied security vulnerabilities residing in HIS networks and the HL7 communication standard, presented a feasible design, and discussed the functions of each component in detail. The main components of a SPHIS should contain monitoring systems, early estimation modules, intrusion detection, network forensics analysis devices, and intrusion response systems. Although there is a gap in current research for developing SPHISs, we surveyed the self-protecting systems and their approaches designed for similar computing systems such as enterprise systems and industrial control systems. We also reviewed and evaluated recent research on each component of the SPHIS. This chapter gives readers an idea how to develop their own self-protecting systems for specific healthcare IT environments to reduce human's burden while enhancing the security posture.

Acknowledgments

This study is supported by supported by the National Science Foundation (NSF) under Grant No. 1812599. Any opinions, findings, and conclusions or recommendations expressed in this material are those of the author and do not necessarily reflect the views of the NSF.

References

[1] I. Masic, The history and new trends of medical informatics, Donald School J. Ultrasound Obstet. Gynecol. 7 (3) (2013) 72–83.

[2] P. Rosenthal, History and evolution of healthcare information systems, 2011. http://instructional1.calstatela.edu/prosent/CIS%20581/chapter4.pptx.

[3] T. Erstad, Analyzing computer based patient records: a review of literature, J. Healthc. Inf. Manag. 17 (4) (2003) 51–57.

[4] J. Henry, Y. Pylypchuk, T. Searcy, V. Patel, Adoption of electronic health record systems among U.S. Non-Federal acute care hospitals: 2008-2015, 2016. http://dashboard.healthit.gov/evaluations/data-briefs/non-federal-acute-care-hospital-ehr-adoption-2008-2015.php.

[5] K. Patel, 6 benefits of IoT for hospitals and healthcare, 2016. http://readwrite.com/2016/07/18/top-6-benefits-internet-things-iot-hospitals-healthcare-facilities-ht1/.

[6] A survey of the cyber security landscape, 2016. http://www-01.ibm.com/common/ssi/cgi-bin/ssialias?htmlfid=SEJ03320USEN.

[7] D. Munro, Data breaches in healthcare totaled over 112 million records in 2015, 2015. http://www.forbes.com/sites/danmunro/2015/12/31/data-breaches-in-healthcare-total-over-112-million-records-in-2015/#2cb37c607fd5.

[8] E. Snell, Healthcare cybersecurity knowledge gaps in phishing awareness, 2016. https://www.wombatsecurity.com/about/news/healthcare-cybersecurity-knowledge-gaps-phishing-awareness.

[9] K. Santos, Phishing attacks target health care sector, 2014. http://idt911.com/education/blog/phishing-attacks-target-health-care-sector.

[10] M.J. Schwartz, Anthem breach: phishing attack cited, 2015. http://www.bankinfosecurity.com/anthem-breach-phishing-attack-cited-a-7895.

[11] 10 largest healthcare cyber attacks of 2015, 2016. http://www.healthdatamanagement.com/slideshow/10-largest-healthcare-cyber-attacks-of-2015#slide-8.

[12] A. Ellison, More than 1,300 dekalb health patients' information compromised by cyberattack, phishing scheme, 2014. http://www.beckershospitalreview.com/healthcare-information-technology/1-361-dekalb-health-patients-information-compromised-by-cyberattack-phishing-scheme.html.

[13] S. Sjouwerman, CEO fraud phishing attack steals 11,000 W-2s from health care workers, 2016. https://blog.knowbe4.com/ceo-fraud-phishing-attack-steals-11000-w-2s-from-health-care-workers.

[14] L. Washburn, St. Joseph's healthcare system falls victim to phishing scam, 2016. http://www.northjersey.com/news/st-joseph-s-healthcare-system-falls-victim-to-phishing-scam-1.1516381.

[15] D. Bisson, Healthcare industry is four times more likely to be impacted by advanced malware than other industries, 2015. http://www.tripwire.com/state-of-security/latest-security-news/healthcare-industry-is-four-times-more-likely-to-be-impacted-by-advanced-malware-than-other-industries/.

[16] R. Moskovitch, N. Nissim, Y. Elovici, Malicious code detection using active learning. in: F. Bonchi, E. Ferrari, W. Jiang, B. Malin (Eds.), Privacy, Security, and Trust in KDD, Springer-Verlag, Berlin, Heidelberg, ISBN: 978-3-642-01717-9, 2009, pp. 74–91, https://doi.org/10.1007/978-3-642-01718-6_6. Malicious code detection using active learning.

[17] R. Winton, Hollywood hospital pays $17,000 in bitcoin to hackers; FBI investigating, 2016. http://www.latimes.com/business/technology/la-me-ln-hollywood-hospital-bitcoin-20160217-story.html.

[18] M.K. McGee, More breaches expose mental health, substance abuse data, 2016. http://www.careersinfosecurity.com/more-breaches-expose-mental-health-substance-abuse-data-a-9390.

[19] Anonymous published more than 10 millions turkish medical healthcare records, 2016. https://www.cyberguerrilla.org/blog/anonymous-published-more-than-10-millions-turkish-medical-healthcare-records/.

[20] Healthcare records from 33 turkish hospitals leaked by anonymous, 2016. https://codingsec.net/2016/05/healthcare-records-33-turkish-hospitals-leaked-online-anonymous/.

[21] J.S. Davis, York hospital breach compromises PII of 1,400 employees, 2016. http://www.scmagazine.com/york-hospital-breach-compromises-pii-of-1400-employees/article/479549/.

[22] Worldwide hospital cyber-attacks posing danger to patient data, 2016. https://security.radware.com/ddos-threats-attacks/threat-advisories-attack-reports/attacks-against-medical-institutions/.

[23] Hacking healthcare IT in 2016: Lessons the Healthcare Industry Can Learn from the OPM Breach, Institute for Critical Infrastructure Technology, 2016.

[24] J. Belliveau, How DDoS attack increase may affect healthcare cybersecurity, 2016. http://www.healthcareitnews.com/news/10-trends-cyberattacks-healthcare-other-industries-new-survey-shows.

[25] DDoS case study: DDoS attack mitigation boston children's hospital, 2016. https://security.radware.com/ddos-experts-insider/ert-case-studies/boston-childrens-hospital-ddos-mitigation-case-study/.

[26] Critical infrastructure sectors — homeland security, 2016. https://www.dhs.gov/critical-infrastructure-sectors, accessed on 05.12.2016.

[27] M. Heller, Health care's cyber-security spend found lacking, 2015. http://ww2.cfo.com/cyber-security-technology/2015/09/health-cares-cyber-security-spend-found-lacking/. accessed on 05.15.2016.

[28] B. Glick, Healthcare sector 340% more prone to IT security threats, 2015. http://www.computerweekly.com/news/4500254005/Healthcare-sector-340-more-prone-to-IT-security-threats. accessed on 03.09.2016.

[29] G. Bell, M. Ebert, HEALTH care and cyber security: increasing threats require increased capabilities, KPMG (2015). https://www.kpmg.com/LU/en/IssuesAndInsights/Articlespublications/Documents/cyber-health-care-survey-kpmg-2015.pdf. accessed on 02.29.2016.

[30] HIMSS cybersecurity survey, 2015. http://www.himss.org/2015-cybersecurity-survey/full-report, accessed on 03.08.2016.

[31] Q. Chen, S. Abdelwahed, A. Erradi, An autonomic detection and protection system for denial of service attack, IASTED/ACTA Press, Las Vegas, NV, USA, 2012.

[32] Q. Chen, S. Abdelwahed, A model-based approach to self-protection in SCADA systems, 9th International Workshop on Feedback Computing, USENIX Association, 2014. https://www.usenix.org/conference/feedbackcomputing14/workshop-program/presentation/chen.

[33] Q. Chen, S. Abdelwahed, A. Erradi, A model-based validated autonomic approach to self-protect computing systems. IEEE Internet Things J. 1 (5) (2014) 446–460. ISSN: 2327-4662, https://doi.org/10.1109/JIOT.2014.2349899.

[34] Q. Chen, S. Abdelwahed, A. Erradi, A model-based approach to self-protection in computing system, in: Proceedings of the 2013 ACM Cloud and Autonomic Computing Conference, 2013.

[35] Q. Chen, S. Abdelwahed, W. Monceaux, Towards automatic security management: a model-based approach, in: Proceedings of the Eighth Annual Cyber Security and Information Intelligence Research Workshop, 2013.

[36] Q. Chen, S. Abdelwahed, Towards realizing self-protecting SCADA systems, Proceedings of the 9th Annual Cyber and Information Security Research Conference, 2014. ISBN: 978-1-4503-2812-8. https://doi.org/10.1145/2602087.2602113.

[37] J.O. Kephart, Autonomic computing: the first decade. Proceedings of the 8th ACM International Conference on Autonomic Computing, ACM, New York, NY, USA, ISBN: 978-1-4503-0607-2. pp. 1–2, https://doi.org/10.1145/1998582.1998584.

[38] M. Parashar, S. Hariri, Autonomic computing: an overview. in: J.-P. Banâtre, P. Fradet, J.-L. Giavitto, O. Michel (Eds.), Unconventional Programming Paradigms: International Workshop UPP 2004, Le Mont Saint Michel, France, September 15–17, 2004, Revised Selected and Invited Papers, Springer Berlin Heidelberg, Berlin, Heidelberg, ISBN: 978-3-540-31482-0, 2005, pp. 257–269, https://doi.org/10.1007/11527800_20.

[39] J.O. Kephart, D.M. Chess, The vision of autonomic computing. Computer 36 (1) (2003) 41–50. ISSN: 0018-9162, https://doi.org/10.1109/MC.2003.1160055.

[40] Y. Diao, J.L. Hellerstein, S. Parekh, R. Griffith, G.E. Kaiser, D. Phung, A control theory foundation for self-managing computing systems. IEEE J. Sel. Areas Commun. 23 (12) (2005) 2213–2222. ISSN: 0733-8716, https://doi.org/10.1109/JSAC.2005.857206.

[41] Q. Chen, J. Lambright, S. Abdelwahed, Towards autonomic security management of healthcare information systems, in: 2016 IEEE First International Conference on Connected Health: Applications, Systems and Engineering Technologies (CHASE), 2016, pp. 113–118.

[42] Q. Chen, J. Lambright, Towards realizing a self-protecting healthcare information system, 2016 IEEE 40th Annual Computer Software and Applications Conference (COMPSAC), 1 2016, pp. 687–690, https://doi.org/10.1109/COMPSAC.2016.264.

[43] Risk Analysis Guide for HITRUST Organizations & Assessors, 2016.

[44] Q. Chen, M. Trivedi, S. Abdelwahed, T. Morris, F.T. Sheldon, Model-based autonomic security management for cyber-physical infrastructures, Int. J. Crit. Infrastruct. 12 (4) (2016) 273.

[45] Architecture of health IT, 2016. https://healthit.ahrq.gov/key-topics/architecture-health-it.

[46] Introduction to HL7 standards, 2016. http://www.hl7.org/implement/standards/, accessed on 03.10.2016.

[47] DICOM: about DICOM, 2016. http://dicom.nema.org/Dicom/about-DICOM.html, accessed on 03.10.2016.

[48] NCPDP terminology, 2016. http://www.cancer.gov/research/resources/terminology/ncpdp, accessed on 03.10.2016.

[49] HL7 ADT-admit discharge transfer, 2016. https://corepointhealth.com/resource-center/hl7-resources/hl7-adt, accessed on 03.10.2016.

[50] G.M. Wood, HL7 membership & basic overview, 2016. https://www.hl7.org/documentcenter/public_temp_4DAF26CA-1C23-BA17-0C75698B827AA7DF/calendarofevents/himss/2016/2016%20HIMSS%20HL7%20Basic%20Overview%20Grant%20Wood.pdf, accessed on 05.11.2016.

[51] G.M. Wood, HL7 and meaningful use, 2012. https://www.hl7.org/documentcenter/public_temp_786B167F-1C23-BA17-0C31D5707666788B/calendarofevents/himss/2012/HL7%20and%20Meaningful%20Use.pdf, accessed on 05.12.2016.

[52] 2016 Interoperability Standards Advisory: Best Available Standards and Implementation Specifications, Office of the National Coordinator for Health IT, 2016. https://www.healthit.gov/sites/default/files/2016-interoperability-standards-advisory-final-508.pdf.

[53] MSH-message header (HL7 V2.5.1), 2016. http://hl7-definition.caristix.com:9010/Default.aspx?version=HL7+v2.5.1&dataType=ID, accessed on 05.12.2016.

[54] B. Claudel, N. De Palma, R. Lachaize, D. Hagimont, Self-protection for distributed component-based applications, in: SSS'06, Berlin, Heidelberg, 2006, pp. 184–198.

[55] G. Heieh, Towards self-protecting security for e-Health CDA documents, 2012.

[56] S. Almomen, A Self-managed Healthcare Emergency Department System (Ph.D. thesis), 2012.

[57] E. Lupu, N. Dulay, M. Sloman, J. Sventek, S. Heeps, S. Strowes, K. Twidle, S.L. Keoh, A. Schaeffer-Filho, AMUSE: autonomic management of ubiquitous e-health systems, Concurrency and Computation: Practice and Experience 20 (3) (2008) 277–295. ISSN: 1532-0634, https://doi.org/10.1002/cpe.1194.

[58] N.K. Salih, T. Zang, Autonomic and cloud computing: management services for healthcare. in: 2012 IEEE Symposium on Industrial Electronics and Applications (ISIEA), 2012, pp. 23–28, https://doi.org/10.1109/ISIEA.2012.6496634.

[59] S. Pandey, W. Voorsluys, S. Niu, A. Khandoker, R. Buyya, An autonomic cloud environment for hosting ECG data analysis services, Future Gener. Comput. Syst. 28 (1) (2012) 147–154. ISSN: 0167-739X, https://doi.org/10.1016/j.future.2011.04.022.

[60] A. Sawand, S. Djahel, Z. Zhang, F. Naït-Abdesselam, Multidisciplinary approaches to achieving efficient and trustworthy eHealth monitoring systems. in: 2014 IEEE/CIC International Conference on Communications in China (ICCC), ISSN 2377-8644, 2014, pp. 187–192, https://doi.org/10.1109/ICCChina.2014.7008269.

[61] R.-T. Liu, N.-F. Huang, C.-N. Kao, C.-H. Chen, C.-C. Chou, A fast pattern-match engine for network processor-based network intrusion detection system, Proceedings ITCC 2004 International Conference on Information Technology: Coding and Computing. vol. 1, 2004, pp. 97–101, https://doi.org/10.1109/ITCC.2004.1286432.

[62] Y.-H. Choi, M.-Y. Jung, S.-W. Seo, L+1-MWM: a fast pattern matching algorithm for high-speed packet filtering, INFOCOM 2008. The 27th Conference on Computer Communications. IEEE, ISSN 0743-166X, 2008, pp. 2288–2296, https://doi.org/10.1109/INFOCOM.2008.297.

[63] M. Fisk, G. Varghese, Applying fast string matching to intrusion detection, Los Alamos National Laboratory, University of California San Diego, 2004, http://woozle.org/mfisk/papers/setmatch-raid.pdf. tech. rep.

[64] T. Jianlong, L. Xia, L. Yanbing, L. Ping, Speeding up pattern matching by optimal partial string extraction. 2011 IEEE Conference on Computer Communications Workshops (INFOCOM WKSHPS), 2011, pp. 1030–1035, https://doi.org/10.1109/INFCOMW.2011.5928778.

[65] G. Vasiliadis, S. Antonatos, M. Polychronakis, E. Markatos, S. Ioannidis, Gnort: high performance network intrusion detection using graphics processors. in: R. Lippmann, E. Kirda, A. Trachtenberg (Eds.), Recent Advances in Intrusion Detection, Lecture Notes in Computer Science, vol. 5230, Springer, Berlin, Heidelberg, ISBN: 978-3-540-87402-7, 2008, pp. 116–134, https://doi.org/10.1007/978-3-540-87403-4_7.

[66] A. Wailly, M. Lacoste, H. Debar, VESPA: multi-layered self-protection for cloud resources, in: ICAC '12, San Jose, California, USA, 2012.

[67] D. Bolzoni, S. Etalle, P. Hartel, Panacea: automating attack classification for anomaly-based network intrusion detection systems, in: E. Kirda, S. Jha, D. Balzarotti (Eds.), Recent Advances in Intrusion Detection, Lecture Notes in Computer Science, vol. 5758, Springer, Berlin, Heidelberg, ISBN: 978-3-642-04341-3, 2009, pp. 1–20.

[68] R. Mitchell, I.R. Chen, Behavior rule specification-based intrusion detection for safety critical medical cyber physical systems, IEEE Trans. Dependable Secure Comput. 12 (1) (2015) 16–30. ISSN: 1545-5971, https://doi.org/10.1109/TDSC.2014.2312327.

[69] N. Boggs, S. Hiremagalore, A. Stavrou, S. Stolfo, Cross-domain collaborative anomaly detection: so far yet so close, in: R. Sommer, D. Balzarotti, G. Maier (Eds.), Recent Advances in Intrusion Detection, Lecture Notes in Computer Science, vol. 6961, Springer, Berlin, Heidelberg, ISBN: 978-3-642-23643-3, 2011, pp. 142–160.

[70] B. Amann, R. Sommer, A. Sharma, S. Hall, A lone wolf no more: supporting network intrusion detection with real-time intelligence, in: Research in Attacks, Intrusions, and

Defenses, Lecture Notes in Computer Science, vol. 7462, Springer Berlin Heidelberg, ISBN: 978-3-642-33337-8, 2012, pp. 314–333.

[71] M. Danforth, WCIS: a prototype for detecting zero-day attacks in web server requests, in: USENIX LISA'11: 25th Large Installation System Administration Conference, Boston, MA, 2011.

[72] G. Modelo-Howard, S. Bagchi, G. Lebanon, Determining placement of intrusion detectors for a distributed application through bayesian network modeling, in: R. Lippmann, E. Kirda, A. Trachtenberg (Eds.), Recent Advances in Intrusion Detection, Lecture Notes in Computer Science, vol. 5230, Springer Berlin Heidelberg, ISBN: 978-3-540-87402-7, 2008, pp. 271–290.

[73] J. Chu, Z. Ge, R. Huber, P. Ji, J. Yates, Y.-C. Yu, ALERT-ID: analyze logs of the network element in real time for intrusion detection, in: D. Balzarotti, S.J. Stolfo, M. Cova (Eds.), Research in Attacks, Intrusions, and Defenses, Lecture Notes in Computer Science, 7462 Springer Berlin Heidelberg, ISBN: 978-3-642-33337-8, 2012, pp. 294–313.

[74] Z. Yu, J.J.P. Tsai, T. Weigert, An adaptive automatically tuning intrusion detection system. ACM Trans. Auton. Adapt. Syst. 3 (3) (2008) 10:1–10:25. ISSN: 1556-4665, https://doi.org/10.1145/1380422.1380425.

[75] E.S. Pilli, R.C. Joshi, R. Niyogi, Network forensic frameworks: survey and research challenges. Digit. Investig. 7 (1) (2010) 14–27. ISSN: 1742-2876, https://doi.org/10.1016/j.diin.2010.02.003.

[76] N. Stakhanova, S. Basu, J. Wong, A taxonomy of intrusion response systems. Int. J. Inf. Comput. Secur. 1 (1/2) (2007) 169–184. ISSN: 1744-1765, https://doi.org/10.1504/IJICS.2007.012248.

[77] C. Gunter, Detecting roles and anomalies in hospital access audit logs. in: Proceedings of the 2014 Workshop on Cyber Security Analytics, Intelligence and Automation, ACM, New York, NY, USA, ISBN: 978-1-4503-3147-0, 2014. pp. 1–1. https://doi.org/10.1145/2665936.2668879.

[78] D. Wang, T. Li, S. Liu, J. Zhang, C. Liu, Dynamical network forensics based on immune agent. in: Third International Conference on Natural Computation, 2007. ICNC 2007, 3, 2007, pp. 651–656, https://doi.org/10.1109/ICNC.2007.345.

[79] Z. Chen, F. Han, J. Cao, X. Jiang, S. Chen, Cloud computing-based forensic analysis for collaborative network security management system, Tsinghua Sci. Technol. 18 (1) (2013) 40–50, https://doi.org/10.1109/TST.2013.6449406.

[80] J. Therdphapiyanak, K. Piromsopa, Applying hadoop for log analysis toward distributed IDS. in: Proceedings of the 7th International Conference on Ubiquitous Information Management and Communication, ACM, New York, NY, USA, ISBN: 978-1-4503-1958-4, 2013, pp. 3:1–3:6, https://doi.org/10.1145/2448556.2448559.

[81] P. Bhatt, E.T. Yano, Analyzing targeted attacks using hadoop applied to forensic investigation, in: The Eighth International Conference on Forensic Computer Science, 2013. Brasilia, Brazil.

[82] Y. Wang, Y. Xiang, W. Zhou, S. Yu, Generating regular expression signatures for network traffic classification in trusted network management. J. Netw. Comput. Appl. 35 (3) (2012) 992–1000. ISSN: 1084-8045, https://doi.org/10.1016/j.jnca.2011.03.017.

[83] Z. Li, L. Wang, Y. Chen, Z.J. Fu, Network-based and attack resilient length signature generation for zero-day polymorphic worms, 2007. Tech. rep.

[84] D. Brumley, J. Newsome, D. Song, H. Wang, S. Jha, Towards automatic generation of vulnerability-based signatures, 2006 IEEE Symposium on Security and Privacy, ISSN 1081-6011, 2006, pp. 15–16, https://doi.org/10.1109/SP.2006.41.

[85] B. Foo, M.W. Glause, G.M. Howard, Y.-S. Wu, S. Bagchi, E. Spafford, S. Bagch, Intrusion response systems: a survey, in: J. Joshi (Ed.), Information Assurance:

Dependability and Security in Networked Systems, chap. 13. Morgan Kaufmann Publishers Inc., 2008, pp. 377–412.

[86] M. Rostami, W. Burleson, A. Juels, F. Koushanfar, Balancing security and utility in medical devices, 2013 50th ACM/EDAC/IEEE Design Automation Conference (DAC), ISSN 0738-100X, 2013, pp. 1–6, https://doi.org/10.1145/2463209.2488750.

[87] M. Rostami, A. Juels, F. Koushanfar, Heart-to-heart (H2H): authentication for implanted medical devices. in: Proceedings of the 2013 ACM SIGSAC Conference on Computer & Communications SecurityACM, New York, NY, USA, ISBN: 978-1-4503-2477-9pp. 1099–1112, https://doi.org/10.1145/2508859.2516658.

[88] Protecting your website with always on SSL, 2012. http://otalliance.org/resources/AOSSL/OTA_Always-On-SSL-White-Paper.pdf.

[89] R. Oppliger, R. Hauser, D. Basin, SSL/TLS session-aware user authentication: a lightweight alternative to client-side certificates, IEEE Computer 41 (3) (2008) 59–65.

[90] S. Jiang, S. Smith, K. Minami, Securing web servers against insider attack, in: ACSAC 2001, 2001, pp. 265–276.

[91] R. Dhamija, J.D. Tygar, The battle against phishing: dynamic security skins. in: SOUPS '05, Pittsburgh, Pennsylvania, ISBN: 1-59593-178-3, 2005, pp. 77–88, https://doi.org/10.1145/1073001.1073009.

[92] V.P. Surwase, S.R. Durugkar, Dynamic security skins-mutual authentication, Bioinfo Secur. Inform. 1 (2011), 1–5.

[93] OWASP secure coding practices quick reference guide, 2010. https://www.owasp.org/images/0/08/OWASP_SCP_Quick_Reference_Guide_v2.pdf.

[94] SQL injection protection: a guide on how to prevent and stop attacks, 2009. http://searchsecurity.techtarget.com/tutorial/SQL-injection-protection-A-guide-on-how-to-prevent-and-sto-attacks#SQL4.

[95] C. Herley, D.A.F. Florencio, Protecting financial institutions from brute-force attacks, in: SEC, 278 Springer, ISBN: 978-0-387-09698-8, 2008, pp. 681–685.

[96] J. Owens, J. Matthews, A study of passwords and methods used in brute-force SSH attacks, Guenevere, 2008.

[97] J. Richter, Distributed protection against distributed brute force attacks, 2011. http://www.feec.vutbr.cz/EEICT/2011/sbornik/03-Doktorske%20projekty/08-Informacni%20systemy/08-xricht14.pdf.

[98] S. Mudhakar, I. Arun, Y. Jian, L. Ling, Mitigating application-level denial of service attacks on web servers: a client-transparent approach, ACM Trans. Web 2 (3) (2008) 15:1–15:49. ISSN: 1559-1131.

[99] A. Stavrou, D.L. Cook, W.G. Morein, A.D. Keromytis, V. Misra, D. Rubenstein, WebSOS: an overlay-based system for protecting web servers from denial of service attacks, J. Comput. Netw. 48 (2005) 2005.

[100] M.M. Patil, U.L. Kulkarni, Mitigating app-DDoS attacks on web servers, Int. J. Comput. Sci. Telecommun. 2 (5) (2011) 13–18.

[101] Y. Luo, F. Szidarovszky, Y.B. Al-Nashif, S. Hariri, Game theory based network security, J. Inform. Secur. 1 (1) (2010) 41–44.

[102] A.J. Ikuomola, A.S. Sodiya, J.O. Nehinbe, A framework for collaborative, adaptive and cost sensitive intrusion response system, 2010 2nd Computer Science and Electronic Engineering Conference (CEEC), 2010, pp. 1–4, https://doi.org/10.1109/CEEC.2010.5606485.

[103] N. Kheir, N. Cuppens-Boulahia, F. Cuppens, H. Debar, A service dependency model for cost-sensitive intrusion response, in: ESORICS'10, Athens, Greece, 2010.

[104] R. Dewri, N. Poolsappasit, I. Ray, D. Whitley, Optimal security hardening using multi-objective optimization on attack tree models of networks, in: CCS'07, Alexandria, Virginia, USA, 2007.

[105] N. Stakhanova, C. Strasburg, S. Basu, J.S. Wong, Towards cost-sensitive assessment of intrusion response selection, J. Comput. Secur. 20 (2–3) (2012) 169–198.
[106] C. Strasburg, N. Stakhanova, S. Basu, J.S. Wong, Intrusion response cost assessment methodology, in: ASIACCS '09, Sydney, Australia, 2009, pp. 388–391.
[107] C. Strasburg, N. Stakhanova, S. Basu, J.S. Wong, A framework for cost sensitive assessment of intrusion response selection. in: COMPSAC '09. 33rd Annual IEEE International Computer Software and Applications Conference, 2009, ISSN 0730-3157, 1, 2009, pp. 355–360, https://doi.org/10.1109/COMPSAC.2009.54.
[108] S. Iannucci, Q. Chen, S. Abdelwahed, High-performance intrusion response planning on many-core architectures, in: IEEE 25th International Conference on Computer Communication and Networks (ICCCN), 2016, pp. 1–6.
[109] K. Fu, J. Blum, Controlling for cybersecurity risks of medical device software, Commun. ACM 56 (10) (2013) 35–37. ISSN: 0001-0782, https://doi.org/10.1145/2508701. http://doi.acm.org/10.1145/2508701.

About the author

Dr. Guenevere (Qian) Chen is an Assistant Professor with the Department of Electrical and Computer Engineering at the University of Texas at San Antonio (UTSA). Before joining UTSA, Dr. Chen was an Assistant Professor and Coordinator of the Computer Science Technology Program at Savannah State University. She earned her Ph.D. degree in Electrical and Computer Engineering from Mississippi State University in 2014. Dr. Chen's primary research area is autonomic computing and cyber security. She developed an Autonomic Security Management framework that proactively defends the computing systems against known and unknown cyber attacks with little or no human intervention. The ASM framework has been successfully applied to secure distributed systems, industrial control systems (e.g., SCADA), high performance and cloud computing and the Internet of Things (IoT) ecosystems. Her research interests include risk assessment, malware behavioral analysis, early warning intrusion detection and response, and end-to-end security solution development.

CHAPTER FOUR

SSIM and ML based QoE enhancement approach in SDN context

Asma Ben Letaifa

MEDIATRON Lab, SUPCOM, University of Carthage, Tunis, Tunisia

Contents

Advances in Computers, Volume 114
ISSN 0065-2458
https://doi.org/10.1016/bs.adcom.2019.02.004

Abstract

Today, video streaming rose above all other traffic types over the internet. While providing this service with a high quality is the most challenging task, researchers are trying to solve the challenge by giving a more efficient network where congestion, broadband limitations and unsatisfied users are limited. In new multimedia based networks, new challenges move from technology-oriented services to user-oriented services which prove the importance of QoE. Service providers' growth depends nowadays not only on QoS parameters but also on clients' feeling and expectation. That is why; service providers must measure received QoE and this becomes a challenge regarding the evaluation of users' feeling. For users, qualitative perception differs from one user to another and service providers affront difficulties to transform the qualitative values into the quantitative one. The QoE evaluation needs more sophisticated methods to describe the real expectation of users. New multimedia applications provide at any user locations media who are sharing video, and communicating together in virtual network. These applications require providing the best possible QoE to consumers. When it comes to us, we present in this paper a machine learning approach combined with adaptive coding in order to provide a better QoE for video streaming services. This solution will be established using SDN architecture. We can justify this choice because we need a centralized architecture, where the totality of the network is known, to predict its status. So, we will implement a machine learning algorithm in the controller: this algorithm, called ML-based SSIM, will calculate approximately the quality needed for a video to be streamed. Finally, the quality found by the ML-based SSIM Algorithm will be combined with the network situation to choose the right coding. First part of the paper deals with an introduction of QoE requirement, metrics and protocols used especially in streaming services, then we give a complete study around Machine learning algorithms and other fields used in literature to enhance QoE. We define in this paper, at first, QoS and QoE, then the serving environment such as mobile cloud computing and software Defined Network (SDN). Then, we give both objective and subjective metrics, expose mathematical approaches used in modeling, predicting and evaluating QoE. Second, we expose the SSIM approach and explain how our proposed one is based on. The last part of the paper deals with experiments: we describe SDN environment deployment, describe scenarios and finally simulate on SDN emulator some topologies to demonstrate the impact of SDN components helping QoE measurement. At the end, we give the results and values. We highlight the future of our proposition.

1. Introduction and challenges

New services such as data storage or processing are outsourced to the Cloud (e.g., Google Apps, or Cloud gaming or streaming on YouTube), and are making users more and more dependent on the network to keep with their daily activities. A metric, called quality of experience (QoE) is appeared to measure user's satisfaction on services done on the internet networks. QoE is a research paradigm to quantify the users' perception on consumed services. One of the most used services is online video streaming, which, in this last decade, has seen a huge growth. The video traffic on internet represented 66% of all global Internet traffic in 2013, and will be about >79% of all Internet traffic by 2018. The evaluation of QoE is based on metrics such as startup time, average playback bitrate, etc. It's providing a better indication of the satisfaction of the clients whose perception is subjective. They have to always consume reachable and react fast services, visual artifacts should not affect their video streams and audio impairments haven't to affect voice calls. Network optimization was traditionally focused on optimizing network properties such as QoS parameters, but these years, QoE depends on both users' perception and quality consumed service, they form end–to–end metrics. We argue that QoE is the measure that is relevant for network operators and service providers.

At the other hand, for current networks, especially mobile social networks (MSNs), providers face a new challenge to provide mobile social user with a better QoE. With new networks generations like 5G, the next network architecture beyond the current 4G, operators and service providers aim to offer low latency connectivity, enable smart and programmable nodes for the Internet of Everything, and adopt the new functions of Software Defined Networking (SDN) and Network Function Virtualization (NFV). The main SDN concept [1] is that both control and data functions are decoupled. New research studies are discussing various challenges posed by SDN, such as virtualization and cloud service applications, networking with other platforms, design of switches and controller, etc. SDN has the potential to facilitate the deployment and management of network applications and services with greater efficiency and beneficiate from a new centralized control mechanism.

Multimedia services know a rapid growth today and become the principal traffic source on the internet. Traditionally, service providers and network operators were facing the challenges to have a higher bandwidth and satisfy multimedia traffic Quality of Service (QoS) requirements. But nowadays, those requirements are fast changing and the actors (providers and operators) are facing new challenges about the quality of audio and on real-time multimedia applications, like video streaming ones. Recently, with the variation of multimedia applications, and in order to guarantee the best QoE to their clients, service providers and operators see emerging new networks architectures and management policies. With those new multimedia applications, new networks called cloud and mobile cloud computing are emerging and represent an efficient environment to provide mobile services for daily life. QoE on mobile devices plays an important role in Mobile Cloud computing MCC and new network topologies, than the other devices properties such as limited hardware resources, mobility, etc. MCC is so different from cloud computing. We argue also that new architectures are defined and we encounter NFV and SDN context, which came with virtualization paradigm and offer a good QoS/QoE with this new concept.

On the other hand, the QoS is defined as a "metric represented by some objective service parameters that can be effectively measured based on the traffic analysis (such as time delay, packet loss rate, etc.)." Moreover, when dealing with Mobile Cloud, SDN networks or IoT environment, a variety of services is considered and just QoS doesn't help enough to satisfy traffic analysis and we need to express a metric called QoE to evaluate the quality. That is why; the QoE metrics is more involved with multimedia services, since it measures the user expectation. The last years, video streaming services represent the main source of generated traffic over the internet. As we have mentioned, although the QoE is used to measure the end-user satisfaction; the problem of its evaluation is much more complicated than the QoS evaluating problem already done in traditional networks with usually objective methods. In other words, in order to evaluate QoE, it is necessary to know how much an end-user would like/dislike the delivered service. Thus, the problem arises of modeling the user perception in machine learning field or mathematical area where both objective and subjective methods are used for this purpose.

When it comes to us, we present in this paper a machine learning approach combined with adaptive coding in order to provide a better QoE for video streaming services. This solution will be established using SDN architecture. This paper deals with an analysis of QoS and QoE challenges [2–5]. Hence, we start by providing an overview of new networks topologies and QoS/QoE definitions and metrics (both objective and subjective), so we focus on extracting relevant key performance indicators for QoE evaluation, then we study mathematical background helping to model QoE especially in the context of Video streaming services. In this paper, we describe at first the study context, we will introduce QoS, QoE, SDN networks, and machine learning fields, and then we describe SDN and make a large overview on QoE metrics and streaming protocols. After that, we present our methodology for service quality estimation and validate it with some experiments: that is why we study different topologies made on SDN networks, some scenarios helping to offer best QoE. The main goal of those scenarios is to analyze the impact of network conditions on video streaming services. Next, we expose our approach to enhance QoE in SDN context with a modified approach based on SSIM metric, we argue our choice and give results of experiments done on streaming service video. At the end of this paper, we conclude the work by giving conclusions, future work and perspectives.

2. About QoS/QoE parameters

Operators and Service Providers must manage the quality to offer service guarantees to their end-users. QoS and QoE represent metrics used in quality of service evaluation. Traditional metrics are already used to estimate the new user QoE when measurements are perceived, by the end-user. Furthermore, QoE measurement is too long to be collected by network measurements. The challenge is to obtain real time QoE measurements at any network node and then to proceed to real time traffic management process.

2.1 QoS

According to the ITU, QoS is defined as: "The totality of characteristics of a telecommunications service that bear on its ability to satisfy stated and implied needs of the user of the service." Delay and percentage of lost packets known as traditional parameters are helping to evaluate the QoS. But we consider that those parameters don't reflect the expectation client. That is why; QoE is more used recently because it reflects really the user's satisfaction degree.

2.2 Intrinsic QoS

Objective parameters such as delay, jitter, and loss are making Intrinsic QoS which represents the network-oriented QoS and constitute the major problem of services providers and operators [6]. Latency or delay, defined as an amount of time that a data packet needs to transfer between two endpoints, jitter, defined as an undesired deviation of latency between two consecutive packets, packet loss, (or lost packets percentage during transmission, bandwidth or amount of available bandwidth represent the main four parameters evaluating the QoS over the network [7]. According to recommendations in Ref. [8], these QoS parameters can be examined from three points of view (1) QoS of the client, (2) QoS that provider is offering, (3) QoS that a client perceives. However, we define Perceived QoS as the quality perceived by the users.

2.3 QoE

QoE is defined in Ref. [8] as "the overall acceptability of an application or service, as perceived subjectively by the end-user." Recommendation tells also that "it includes the complete end-to-end information about client, terminal, network infrastructure; and may be influenced by the client context." Hence, measuring a subjective QoE may differ from one client to another, it is usually estimated using objective parameters. The relationship between QoE and QoS is non-trivial and we have to analyze if additional factors can influence the perception of quality for delivery of multimedia content especially in cloud computing environment and SDN networks. Fig. 1 shows factors contributing to QoE.

Subjective QoE metrics quantify the perceived quality of a service by the end-users. Hence, QoE needs to gather numerous parameters such as the encoding, transport, content, type of terminal, network, services infrastructure, media encoding, etc. as well as the user's expectations [9]. Some major multi-dimensions like effectiveness, usability, efficiency, expectations and

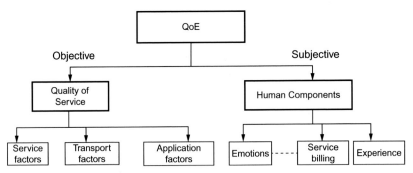

Fig. 1 QoE contributing factors.

context are also affecting QoE. Moreover, those subjective QoE concepts are expensive and time-consuming since it requires human participation. Authors of Refs. [10,11] tell us that MOS is representing the QoE but says also that QoE assessment can be subjective (qualitative) and objective (quantitative). Subjective methods are based on people participation, when objective methods based on several parameters measurement.

2.3.1 Objective parameters

For quantitative QoE measurement, we can found three models at literature: No-reference (NR), Reduced-reference (RR) and Full-reference (FR). Table 1 summarizes most popular objective video quality estimation methods in this context.

2.3.2 Subjective parameters

Table 2 gives a survey to main subjective parameters for both speech and multimedia traffic quality measurements.

By providing mathematical models for the quality estimation, Objective methods try to overcome limitations caused by subjective ones, but known models do not consider yet other context parameters (terminal, location, time-of-the day, etc.). That is why, we think that we have to propose a

Table 1 Objective video quality methods.

NON Human Visual System (HVS) perception	Peak Signal-to-Noise Ratio (PSNR)	Poorly correlates with the human perception of visual quality
	Video Quality Model (VQM)	Perceptual effects of video impairments. A single metric that combines blurring, jerky/ unnatural motion, global noise, block and color distortion
	Perceptual Video Quality-of-Experience Measurement (PEVQ)	Degraded video signal is compared to the original signal on a perceptual basis
Human Visual System (HVS)	Moving Picture Quality Metric (MPQM), Motion based Video Integrity Evaluation (MOVIE), Perceptual Quality Index (PQI)	

Table 2 Subjective parameters.

Mean opinion score (MOS) [12]	For speech quality and multimedia traffic average human rating (say 1–5 scale)
Double Stimulus Continuous Quality Scale (DSCQS) [13] Double Stimulus Impairment Scale (DSIS) Single Stimulus Continuous Quality Scale (SSCQS) Single Stimulus Continuous Quality Evaluation (SSCQE)	Subjective video quality evaluation
Absolute Category Rating (ACR)	Video quality instantaneously
Stimulus Comparison (SC)	

hybrid method, like what is done in INRIA with PSQA model and insert parameters such as environmental, personal, social, cultural, technological and organizational class.

2.4 User experience

According to Mayer [12], the user experience is classified thank to four measures: perception, rendering quality, physiological, and psychological. According to authors of Ref. [14], for tele-immersive services, no prediction is available, such as World Opera application or collaborative gaming. However, prediction models exist for interactive multimedia services such as VoIP. The evaluation of objective QoE is strongly related to the network, the QoS metrics are really used to calculate this QoE value. Difference between both evaluation methods (about network/human) may cause that providing the best QoS do not necessarily provide the best QoE.

2.5 Correlation between QoS parameters and QoE

QoS metrics are generally related to device including CPU, memory use, packet loss, delay or jitter. When QoS focuses on the equipment's controllers or network traffic, QoE performance indicators are user-centric (for instance: webpage loading time, application response duration, VoIP MOS) and reflect the user's satisfaction and feelings. The relationship between QoS and QoE is not linear so correlation is hard to estimate. We can find in literature a variety of QoS/QoE correlation algorithms. In Ref. [14], we describe, for example, the relationship between the QoS and QoE metrics, including media type or another algorithm that describes QoE evaluation algorithm adapted to a human's brain in order to "predict" number of like/dislike. Some works found in literature describe also several QoS metrics that impact overall QoE: We can simply predict QoE, with appropriate assumptions based on QoS measurement or we just deduce the net required service layer performance based on appropriate QoS assumptions. Several correlation models for Video on Demand (VoD) service or web video estimation models are studied.

3. Environment and challenges
3.1 SDN: Software defined networking

SDN architecture presents a new solution that consists of separating the control plane from the data plane which is typically coupled together. Network functions traditionally realized in specific hardware can now be abstracted and virtualized on any equipment. A split between control and data path nodes is performed, so a centralized controller has a global view of the network while the data plane includes devices which simply forward packets following rules expressed by the controller. In order to communicate between these two layers, an open standard protocol is employed. This separation between the two layers simplifies the network management and help to simply program network control.

The main concept of SDN architecture consists on the separation between control and forwarding functions. The Figs. 1 and 2 demonstrate the multiple components of an SDN architecture which is based on three layers separated by open interfaces. The *SDN application plane* is a layer composed of a variety of applications that communicate via Northbound APIs. It's responsible for management, reporting functionalities such as monitoring or security. The *SDN controller* plane represents the main entity in the network that facilitates the creation/destruction of network paths. Typical SDN Controllers are OpenDaylight and Floodlight. The *SDN data plane* includes different devices deprived from any intelligence. They simply execute the controller's rules. The SDN architecture defines also the key interfaces between the different layers, which are *East/West bound API* that are implemented by the different controllers of the SDN and used to facilitate communications between them. Hyperflow is one representative example of such APIs. *Southbound API* is implemented by the different forwarding devices in the SDN and enabling the communication between these devices and the controllers. For such APIs, we can enumerate OpenFlow or NetConf.

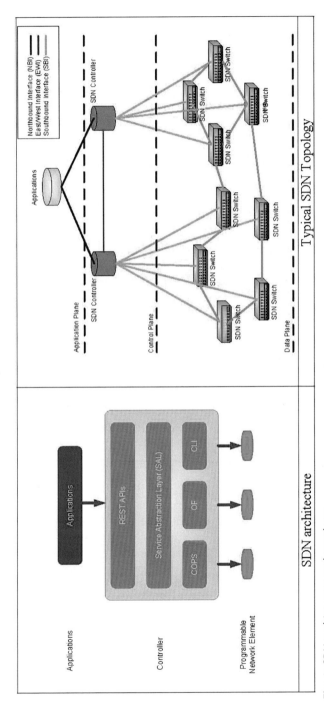

Fig. 2 SDN architecture and topology.

The *Northbound API* is Implemented by the controllers of the SDN and used to facilitate the communication between controllers and the network management applications. In such SDN architecture, enhancing the QoE became easier by implementing specific algorithms within the controller entity. The SDN manages in such case to enable new functionalities such as QoE monitoring and enforcement functions.

3.2 QoS and QoE context

Quality of Service is defined by the ITU as the "totality of characteristics of an entity that bear on its ability to satisfy stated and implied needs." It defines also other metrics helping to providing guarantees in term of delay, jitter, and packet loss that are useful for user satisfaction.

For example, ITU-T Recommendation G.1010 states that for interactive voice communication the delay should be below 150 ms and the packet loss rate should be below 3%. In other hand, QoE is defined as "the degree of delight or annoyance of the user of an application or service. It results from the fulfillment user expectations with respect to the utility/enjoyment of the application or service in the light of the users' personality and current state." QoE aims to capture the users' perception when using a service. To address this challenge, we review in this paper the most common quality metrics especially for video streaming services. These metrics estimate a set of influence factors impacting QoE. We illustrate in Table 3 an overview of relevant metrics, a selection of what we consider to be the most relevant and often used QoE metrics for speech, video, and web browsing. For an

Table 3 Overview of QoE metrics.

	Speech	Video
No-Reference NR	E-Model	Annoyance prediction Frame freeze Loss Visibility Blackness/blurriness PSQA: Pseudo-Subjective Quality Assessment
Reduced-reference RR		Webster algorithm LHS: Local Harmonic Strength
Full-Reference FR	PESQ PQLA	MSE: Mean Square Error PSNR: Peak Signal-to-Noise Ratio SSIM: Structural SIMilarity VQM: Video Quality Metric

extended overview of available metrics, we refer to Ref. [15] for speech quality, to Ref. [16] for image quality, and to Ref. [17] for video quality. QoE metrics can also be classified into three categories by the required amount of reference information. Full-reference FR metrics estimate the QoE score based on the original signal and received perturbed signal. No-reference NR metrics account for this challenge and estimate QoE_score purely based on the received signal. Reduced-reference RR metrics account for this inaccuracy and estimate it based on a subset of features.

The challenge on QoE is encountered since the users' perception is subjective. The main objective of QoE is to quantify the users' perception of the applications from service generation, including transport entities until the end device's screen or audio unit. Both User satisfaction and service qualities are strongly correlated from end-to-end serving entities.

3.3 How can SDN controllers manage the QoS/QoE?

SDN allows the evolution of mechanism and service deployment on programmable networks. Hardware and control unit are decoupled in SDN. Network intelligence is centralized in controllers and the other nodes are reduced to simple flow forwarding entities that can be programmed via interfaces. Openflow is the standard protocol that acts as interface between the Network Controller and the network entities [18]. Some related works that offer QoE monitoring done all over SDN are described in Ref. [19]. Other papers focus on the design of video control planes and consider the use of resource allocation based on video quality [20] or QoE [21]. Since the QoE is not directly measurable, network operators face the challenges to calculate a QoE Key Performance Indicator KPI value from measurable QoS parameters. We are facing three assumptions: (1) the network controller has a global view of the network, (2) it monitors others network nodes, (3) openFlow does not support QoS configuration, we argue that we can easily implement a new rule or a new module in the controller node and implement it to enhance QoE. For example, authors of Ref. [18] propose for this case two approaches: (i) the first one by allocating network bandwidth slices to video flows and (ii) the second one by guiding video players in the video bitrate selection. They implement the open Day light controller through two communicating HTTP servers, one hosted by the controller machine and one hosted by the switch. The HTTP server hosted at the controller maintains the Active Flows Table, executes the Optimization Module in a Python thread, and establishes communication pipes; they prove that several QoE metrics, such as Video Quality Fairness, video quality and switching frequency can affect SDN performance. In another work [1,22], authors used RYU controller to receive per-client QoS policy messages coming from the application layer. The SDN-controller here passes these messages to the internal SDN-based application (implemented inside the RYU controller responsible for resource allocation, monitoring and QoS provisioning) to configure, allocate, provision, and monitor the network resources for each client accordingly. In different framework Sangeeta et al. [23] proposed a model where all the intelligence and the administration of the work are located in the SDN network, their ultimate goal is to optimize the overall QoE. To estimate the available bandwidth correctly, the proposed optimization application is based on several metrics such as device type the network topology which are obtained from the SDN controller swiftly and easily.

In one word, in all these research works, the role of the controller is clear and very important where it facilitates the communication, the management and the administration of the whole architecture, and the addition of the new rules of QoE optimization.

To fast deploy and test new services, applications and protocols, we can test scenarios on an emulation and simulation tools named Mininet. We can also use ns-3 network simulator which supports OpenFlow switches.

3.4 QoE challenges in SDN environment

We find in literature some proposals answering this QoE challenges. SDN/NFV based solution for QoE monitoring in mobile networks developed in "ISAAR project" proposes in this purpose three functional components: QoE Monitoring (QMON) for flow detection and assessment, QoE Rules (QRULE) for policy rules and permission checking and finally QoE Enforcement (QEN) for respective flow manipulation. We argue that SDN/NFV augments the existing QoE monitoring and enforcement. That is why described project ISAAR makes use of SDN to selectively copy out flows as well as to enforce flow manipulation actions. It goes throw functional block structure (QMON, QRULE, QEN) allowing direct NFV implementation. Those three function split options have been presented: SDN only, SDN + NFV "light" and SDN + NFV "full" with good QoE results. On the other hand, paper [24] introduces an in-network QoE Measurement Framework (IQMF) that provides live network-wide QoE measurements. The framework manages the network without any interaction with client. For such purpose, IQMF adopts quantitative metrics to measure the video fidelity QoE and representation impact.

Several researchers have investigated Mobile context with SDN. Zafar et al. [25] focus on mobile application detection. They use a ML–based traffic classification method in order to identify flow types in SDN. Ali et al. [26] used OpenFlow to cope with MPLS-TE and MPLS VPNs. Saurav et al. [27] argue aggregation engineering in a packet circuit network [28]. Michael et al. [29] have investigated on YouTube application.

Operators using 3GPP QoS mechanism address service flow management. An overview streaming application based on HTTP is found in Ref. [30]. Other approaches also address applications enhancements. They define HTTP Adaptive Streaming Services (HAS) [31] as a new way to keep with the video streaming quality. Other approaches are about sharing contents on Fixed-Mobile Convergence (FMC) [32]. The ISAAR framework presented in Ref. [33] follows a different approach. It aims for service flow differentiation without PCRF support or PCRF based flow treatment.

Video streaming services have grown to become increasingly popular over the recent years. Enhancing video delivery and clients QoE, Kim and Choi [34] proposed the Server and Network Assisted DASH (SAND) architecture. SAND is a control plane for video delivery that obtains QoE metrics from the users (clients) and returns network-based measurements to help the clients enhance their overall QoE. The authors of Ref. [35] propose a measurement approach for YouTube video traffic. QMON [36] enhances previous work by supporting more video codecs. Specifically, it estimates the playout buffer levels [37] and the number of stalls and their duration, and then calculates the MOS (Mean Opinion Score) for YouTube videos.

3.5 MSN, CDN and content-centric networks in big data context

We assist nowadays to the explosion of mobile social networks (MSNs) which provide users with various social applications who have disposition to obtain data on mobile devices instead of their traditional desktop computers. In this context, Mobile Data are delivered over Content-Centric Networks CCNs [38] to settle new challenges (1) we can store a very large volume of mobile big data by using the cache space in the content store, (2) we can manage fast mobile social data exchange by using named data and interest packets, (3) we can deliver variable mobile big data with corresponding agents nodes selection. To face with those challenges, future works have to answer these three questions: (1) How to control dynamic mobile big data? (2) How to make an efficient resource allocation for mobile big data? (3) How to guarantee privacy in case of multiple replicas?

CCN or differently named Information-centric networks ICN allow the network to cache data by attaching storage to routers. Recent works consider different aspects of video streaming and rate adaptation. The work of Adhikari et al. described in Ref. [39] is performing measurements to evaluate Netflix. The authors of papers [16,40] study dynamic rate adaptation for streaming services and facilitate the distribution of video. A new field is appeared these last years: Big data in MSNs which can be described as mobile big data, which have large volume, wide variety, fast velocity, and economic value. Different from conventional big data, mobile social big data should have in particular the following characteristics [38] Volume, Velocity, Variety and Value. Within MSNs, mobile big data have a wide variety of data, including video, audio, and file sharing. We give here some statistics taken from Ref. [38]: In 2015, mobile data traffic caused by video is 2,399,765/ mo; mobile data traffic of other file sharing is about 74,694/mo. Besides, compared to 4.2 EB/mo of mobile data traffic in 2015, mobile data traffic is predicted to ranch 24.3 EB/mo in 2019. With this growth, new challenges, with new QoE requirements, are faced: (1) better services are hoped to be provided for mobile social users with a satisfactory (QoE). (2) Storing mobile big data will face new difficulties due to the limited capacity of mobile storage in MSNs. (3) Mobile big data should have a fast rate to match the speed of data production. (4) The variety of mobile big data should be considered to satisfy the demands of different mobile social users. (5) Mobile big data should be studied with consideration of their value. Research studies have to discuss new challenges in MSN: How to store a very large volume of mobile big data in MSNs? How to manage mobile social data with fast velocity? How to deliver mobile big data with variety? How to control mobile big data based on value?

4. QoS and QoE metrics in different services applications

Network, content and video service providers take advantage of content distribution networks (CDNs) and would like to ensure a high degree of video QoE for their clients. We give in Table 4 the most used applications on data networks and explain after the main used metrics for each traffic.

Table 3 provides a brief description for multimedia quality metric as they had been used in literature.

Table 4 Traffic collection.

Traffic class	Video streaming	Video chat/VoIP	P2P torrent	Cloud storage	Online games	Email client
Application	YouTube Netflix Dailymotion	Skype Gtalk Facebook Messenger	VUZE BitTorrent	Dropbox Google Drive OneDrive	8-Ball Pool Treasure Hunt	Thunderbird Outlook

Over the years, multimedia applications have conquered several segments of the telecommunications field. Today, we are dealing with multimedia services in many areas, starting with different digital television systems (e.g., DVB), video telephony, video on demand (VOD), Internet Protocol TV services (IPTV), the voice over IP (VoIP) or simply, video sharing services like YouTube or Dailymotion. The development of these services and the end-to-end optimization of these systems are closely related to the perception of quality by the user and his satisfaction with the service rendered. In this sense, there is a deep need for a measure that reflects the satisfaction and perception of users. Indeed, media service providers are increasingly interested in evaluating the performance of their services provided as perceived by end users, in order to improve and better understand the needs of their customers. Network operators are also interested in this measure to optimize network resources and even (re)configure network settings to increase user satisfaction. There are several ways to get information about perceived quality. On the one hand, there are subjective evaluations done in fully equipped laboratories to investigate the perception of the end user. On the other hand, there are objective quality measures [41], which are often used to study the measurable parameters of the entire system, describing the quality of service (QoS) in a technical way. However, these parameters cannot describe all the variables that influence the perception of quality on the end-user's side. For this reason, a new measure, called "Quality of Experience (QoE)," has been defined to reflect the quality perceived by end users.

5. How modeling QoS/QoE in related work?

To evaluate QoS/QoE, we choose in this article to focus on video streaming services [42–44] and all related requirements for serving this type of application. We define the concept, metrics and KPIs used to model it before giving mathematical models helping to model it using real traffic.

5.1 Video streaming overview

Video streaming playback, real-time video chats applications, online TV have some specific conceptions as media stream is done over IP; Video can be live (rewound/paused/recorded) or on demand (sought/played/paused). This can be done on specific components such as Distributed File Storage, Edge Servers, DRM System and Client Player. As an overview of video streaming techniques and protocols, the most famous are Akamai, Google Global Cache, Limelight Networks where video content is replicated from the original server to several called replica located near end-users especially when we use protocols based on HTTP. At the end, HLS or MPEGDASH known as delivery protocols are supported by video playback devices.

To distribute video on the internet, we find two video delivery methods: progressive download (process based on TCP/FTP or TCP/HTTP) or HTTP Adaptive Streaming (in order to solve video downloading problem). Streaming techniques from literature are described below:

- Progressive Download: where a client have the choice to get the appropriate codecs to play data using HTTP or HTTPS.
- HTTP Live Streaming: MPEG family transport channels offering live and on demand media.

In Transport layer, TCP is the most used protocol that offers "guaranteed delivery." The second one is the unreliable UDP used in real-time operation. However, thanks to bandwidth availability, video is now delivered very efficiently using TCP caches placed the nearest to the client equipment. In Application layer, the most known protocols are HTTP (text/binary data) and RTP (audio/video over IP networks and streaming media).

5.2 Modeling QoE in a video streaming context

As we have discuss, there have been a number of research works done with the objective of video traffic QoE estimation based on media parameters (PSNR, VQM, SSIM, and PEVQ). However, very limited works can be found on QoE estimation of video transmission (Table 5). For other multimedia application over the internet, Various QoE models are developed for specific applications such as web browsing, audio/video services, online gaming and telepresence [45].

In order to evaluate QoE, we can use some QoS parameters such as packet loss, burst loss, jitter, delay, GoP-Group of Picture length, summarized in Table 4 as QoV, QoA and QoS indicators in Table 6.

Table 5 Application.

Applications	Impact of:
Web browsing	Session time and temporal correlations
Audio services	PESQ (Perceptual Evaluation of Speech Quality) delay, echo
Video services	PSNR, MSE: codec type, video frame and packet loss
Online gaming	Quality factors such as interactivity and consistency
Telepresence	Interactivity, consistency and vividness

Table 6 Indicators.

Quality of video indicators (QoV)	Video quality score (VQS)
	Frequency of Quality shifting
Quality of Application indicators (QoA)	Initial buffering time
	Mean buffering duration
	Rebuffering frequency
Quality of Service indicators (QoS)	Bandwidth
	Latency

5.3 Mathematical background

Nowadays, multimedia services of audio, image, video or data traffic knows a fast progress on both public and private networks. The relationship between QoS and QoE relationship is not linear so it difficult to be estimated. Moreover various QoS/QoE algorithms can be found; Table 7 summarizes the mathematical methods that model QoE.

6. Video streaming protocols

Despite scarce resources available, mobile operators are facing the challenge to serve multiple clients while maximizing their QoE. New techniques, called adaptive streaming, are introduced for video transmission over varying channels such as wireless network. In order to match the available channel bandwidth, these techniques vary continuously both transmitted video quality and bitrate. To adaptively select video rates, video clients use those video streaming protocols. Video rate adaptation approaches use two factors to maximize viewer QoE: video quality or video rates and

Table 7 Mathematical background.

Method	Work done with it	How it is working?
Fuzzy logic	Neural network and fuzzy logic formulae [46]	Prediction of the user satisfaction
Rough set theory	Inference rules for fuzzy expert system [47,48]	Subjective data set in the form of a conditional attribute set and a decision attribute set
Fuzzy expert systems	[49]	Making decision with imprecise information
Regression analysis	Web service QoE estimation method [50]	Calculates indexes of a correlation function from the subjective test data; however, the quality estimated by this method has high MOS error margin
Artificial intelligence techniques	Artificial Neural Networks (ANN) [51]	Connects the QoE directly to QoS metrics according to the corresponding level of QoE which was estimated by employing a Multilayer Artificial Neural Network (ANN)
	Random Neural Networks (RNN)[52]	Adjusts the input network parameters in order to get the ideal output and satisfy QoE based on those parameters the source Bit Rate (BR), the Frame Rate (FR), the Packet Loss Rate (PLR), the Consecutively Lost Packet (CLP), and the Ratio of the encoded (RA)
	Adaptive Neural Inference System ANFIS [53]	Learning technique can be found in the work of Khan [39] which presented new models to estimate the video quality based on an Adaptive Neural Inference System (ANFIS) Work was limited to four parameters: FR, SBR, packet error rate (PER), bandwidth
	Machine Learning [19]	Decision trees (DTs), random neural networks (RNNs), hidden Markov models (HMMs), Bayesian networks (BNs) and dynamic Bayesian networks (DBNs)
Learning logic networks	[54]	Evaluating the QoE of web services
Neural networks	PSQA metric [55]	Numerical values such as packet loss percentage and mean loss burst size to output MOS based on RNNs

rebuffering time. There are three mainstream video rate adaptation methods: Network capacity based method, Buffer state based method and Hybrid method based on (1) HTTP caches placed in the network in order to reduce server load, (2) video split in smaller segments called chunks, (3) different video representation to permit users to request the one that matches at best their capabilities. We can give here some details about main streaming protocols.

Table 8 gives QoE metrics and Table 9 gives a survey of streaming protocols used and how they work.

At the end of this survey, we can argue that many multimedia streaming applications are still using HTTP. Using HTTP has some disadvantages: this protocol is at first designed for best effort and not for real-time multimedia transport, and second, it is unable to follow bandwidth variations of IP-based networks. For traditional RTP/UDP streaming, adaptation is difficult, some new studies propose adaptive HTTP streaming to encode content at multiple bit-rates. In wireless context, authors of Ref. [71] propose a schema for multi-user DASH that optimizes the media delivery to multiple clients. Over the past years, authors propose methods for video quality assessment. Most of them [72,73] are measuring a visual quality of video frame called video spatial quality and withdraw the temporal effects. In streaming session [74], some new models are proposed for video temporal quality, but they don't include the variation of bit rate (visual quality). In addition, Jiang et al. [75] proposed FESTIVE a general bitrate adaptation framework that implies a number of methods that fight to achieve a trade-off between video stability, fairness and efficiency. They used a dynamic rate selection heuristic is used to compensate for the biased interaction between the bit rate and the estimated bandwidth, as well as to make a compromise between stability and efficiency using a delayed update approach. But the problem of this model is the instability with the increase of the number of clients in the network and the effect of the bandwidth overestimation. In addition, it is not very sensitive to bandwidth fluctuations, which can lead to a significant undershoot of the buffer. It struggles to ensure fair bandwidth between competing DASH clients in a shared network environment. Tasnim et al. [76] proposed a novel approach based on Machine Learning and adaptive bitrate to enhance the video quality perceived and ameliorate the QoE in SDN context, they used Machine Learning to estimate the best level of the following video segment based on the current network conditions and client requirements (device, video content requested, etc.), in a second step authors proposed an algorithm to select the suitable quality level depending on the

Table 8 QoE metrics in multimedia applications.

	Internet traffic service	VoIP [56]	Video [57]	Video streaming
Specificity	Streaming protocols control the data transfer between the multimedia server and the clients	Comparing performance of various Voice codecs, such as G.711, GSM, iLBC, Speex, and Skype's codec		YouTube monitoring evaluate QoE of MP4, Flash Video (FLV) as well as WebM video in Standard Definition (SD) and High Definition (HD) format
Based on	RTP RTP over UDP for real-time transfers	[24] Subj Quality assessment Opinion rating Opinion equivalent-Q E-Model Speech-layer Obj models PESQ, P.AAM [34] Measured PESQ [58] CAC/CS algorithm [59] adaptive/non adaptive policy [60] adaptive rate control	PSNR (Obj) evaluation of the quality difference among pictures VQM: perceptual effects of video impairments including blurring, unnatural motion, global noise, block distortion color distortion, and combines them into a single metric MPQM: incorporates two human vision characteristics: contrast sensitivity and masking SSIM: is using the structural distortion measurement instead of the error RMSE: average squared error NQM: a degraded image that has been subjective to linear frequency distortion and additive noise injection	HTTP bases streaming uses TCP for reliable data delivery [61] Client based video quality estimation approache, Passive YouTube QoE Monitoring for ISPs approach, Network Monitoring in EPC system QoE indicators of HTTP video streaming: Number of stalling events and their respective lengths, Initial buffering time: denoting the time to fill up the application buffer and to start the playback

Continued

Table 8 QoE metrics in multimedia applications.—cont'd

	Internet traffic service	VoIP [56]	Video [57]	Video streaming
Services	IPTV–VOD	Skype–Viber	Video image	Youtube HTTP-based streaming
Advantages	Small delay		VQM: objective video quality metric, good correlation with human perception MPQM: giving good correlation with subjective tests for some material RMSE and SSIM are the metrics that measure the error or similarity between a referenced image and a fused image	Neither video nor audio suffer from impairments caused by lost data
Disadvantages	UDP unreliable protocol Packet loss in multimedia streams Distortion of content		VQM doesn't consider the impairment due to level variation	Rebuffering events alter the temporal structure of the video and result in stalling events so degrade QoE

Table 9 Video streaming protocols overview.

Protocols	Definition	How does it work?
HTTP Live streaming protocol [62]	Created by Apple to communicate with iOS and Apple TV devices and Macs running OSX	HLS can distribute both live and on-demand files Uniq technology available for adaptively streaming to Apple devices
Adobe's HTTP Dynamic Streaming	HDS enables on-demand and live adaptive bitrates video delivery of standards-based MP4 media over regular HTTP connections	HDS provides tools for integration of content preparation into existing encoding workflows
Microsoft's Silverlight Smooth Streaming		Consumers get minimal buffering and fast start-up time, by adapting the quality of the video stream in real-time based upon the consumer's changing bandwidth and CPU conditions
HTTP Adaptive Streaming Services [30,31,63–65]	HAS: a new way to adapt the video streaming quality based on the observed transport quality	The concept is based on the idea to adapt the bandwidth required by the video stream to the throughput available on the network path from the stream source to the client The adaptation is performed by varying the quality of the streamed video and thus its bit-rate, that is, the number of bits required to encode 1 s of playback. The aim is to divide the video stream into segments and to encode each of the segments in multiple quality levels, called representations. Based on his estimation of the available throughput, a client might request subsequent segments at different quality levels in order to cope with varying network conditions
Jarnikov et al. [66]	Proposed to calculate the adaptation strategy using a Markov decision process	
Akhshabi et al. [67]	Evaluated two major commercial players and one open source player that use the technology of adaptive streaming over HTTP	Three aspects: reaction to persistent or short-term throughput changes, ability of two players to properly operate on a shared network path, and if the player is able to sustain a short playback delay and thus perform well with live content

Continued

Table 9 Video streaming protocols overview.—cont'd

Protocols	Definition	How does it work?
Zink et al. [68]	Investigate the impacts of video quality variations on human perception	
The authors of Ref. [69]	Implemented an algorithm which aims at avoiding interruptions of playback, maximizing video quality, minimizing the number of video quality shifts and minimizing the delay between user's request and the start of the playback	
DASH: Dynamic Adaptive Streaming HTTP	Based on 3GPP's AHS and the Open TV Forum's HAS It specifies use of either fMP4 or transport stream (TS) chunks or an XML manifest DASH is a worldwide standard for adaptive streaming of video, audio and other media such as closed captioning	DASH enables delivering media content from conventional HTTP web servers. DASH works by splitting the media content into a sequence of small segments, encoding each segment into several versions with different bit rates and quality, and streaming the segments according to the requests from mobile client On the mobile device side, the DASH client will keep monitoring the network and dynamically select the suitable version for the next segment that need to be downloaded, depending on the current network conditions Adaptive Video Streaming Measurements: Largescale studies of DASH performance have been conducted in commercial video streaming platforms such as Hulu, Netix and Vudu [31]. Similar measurement studies also covered aspects of the transmission behavior of video clients [67], DASH network traffic characteristics [70], and DASH QoE [45] [71] proposes a QoE-driven multi-user DASH scheme that optimizes the adaptive HTTP media delivery to multiple clients in a wireless cell
SCALABLE VIDEO CODING (SVC)	Least Recently used (LRU), Least Frequently used (LFU) or Chunk-based Caching (CC) are some different cache replacement algorithms [70]	Improves the cache performance based on some especial criteria/metric

estimated best level, the choice is based on the buffer occupancy and the maximum number of intermediate quality video segments to be downloaded. Besides, authors in Ref. [20] presented an SDN-assisted QoE Fairness Framework, called QFF, where they try to enhance the heterogeneous DASH multiuser QoE by taking into account two main constraints, namely, the devices' resolutions and current network requirements. This framework is based on SDN network and the OpenFlow protocol, However, this model neglect the buffer occupancy metric that's why it can't support a large number of clients. In another work [77] authors proposed the Server and Network-assisted DASH (SAND) architecture where they are based on the fact that client-driven rate adaptation provides less control for the network and service providers. This model introduces a centralized control component that offers asynchronous network-to-client and network-to-network communication. It receives QoE metrics from the clients and returns network feedback measurements, which are used by clients heuristically in their adaptation logics. SDNDASH proposed by Abdelhak et al. [22] aims to ease the scalability issues and improve the QoE. This architecture manages and allocates the network resources dynamically for each client based on its expected QoE by taking into account different metrics that affect the QoE (e.g., buffer sizes, display resolutions, quality requirements, available bandwidth, latency, etc.). Experimental results show that the proposed architecture significantly enhances scalability by improving per-client QoE by at least 30% and supporting up to 80% more clients with the same QoE compared to the conventional schemes.

Authors of Refs. [78,79] considered both spatial and temporal quality, and studied the impact of bit rate variation on user experience and developed a quantifiable measure DASH video. We can conclude with an assumption for DASH video, that user experience mainly depends on both subjective (temporal quality-initial delay and spatial quality-level variation) and objective factors (initial delay, stall or frame freezing and level variation).

7. Enhancing QoE approaches

7.1 Machine learning approach

Machine learning (ML) can be defined in Ref. [80] as "A study of making machines acquires new knowledge, new skills and reorganize existing knowledge." ML is used in three approaches: supervised, unsupervised and semi-supervised. Four basic types of learning are known: (1) Classification (or supervised learning), (2) Clustering (or unsupervised learning), (3) Association, (4) Numeric prediction. ML algorithms (Table 10) can help in

Table 10 ML algorithms.

ML-algorithm	How they work?
Resulting decision tree k-nearest neighbors (kNN) algorithm	The k−nearest-neighbor algorithm is a pattern recognition model that can be used for classification and regression. Often abbreviated to k-NN, the k is a positive integer, typically small. In classification or regression, the entry will consist of the k closest training examples in a space. In our work we focus on the k−NN classification. In this method, the output is class membership. This will assign a new object to the most common class among its k nearest neighbors. In the case of k = 1, the object is assigned to the class of the nearest neighbor
Naîve Bayes (NB)	Selects optimal (probabilistic) estimation of precision values based on analysis of training data using Bayes' theorem, assuming highly independent relationship between features
Best-first decision tree (BFTree)	Uses binary splitting for nominal as well as numeric attributes and uses a top-down decision tree derivation approach such that the best split is added at each step
Regression tree representative (REPTree)	For general use, decision trees are used to visually represent decisions and show or inform decision making. When working with machine learning and data mining, decision trees are used as a predictive model. These models map the data observations and draw conclusions about the target value of the data The goal of decision tree is to create a model that will predict the value of a target based on input variables. In the predictive model, the attributes of the data that are determined by the observation are represented by the branches, while the conclusions about the target value of the data are represented in the leaf. When learning a tree, the source data is divided into subsets based on an attribute value test, which is repeated recursively on each of the derived subsets. Once the subset of a node has the value equivalent to its target value, the recursion process will be complete a fast implementation of decision tree learning which builds a decision/regression tree using information gain and variance with reduced–error pruning along with backfitting
Decision tables and naive Bayes (DTNB)	A hybrid classifier which combines decision tables along with naïve Bayes and evaluates the benefit of dividing available features into disjoint sets to be used by each algorithm, respectively
Bayesian network (BayesNet)	An acyclic directed graph that represents a set of features as its vertices and the probabilistic relationship among features as graph edges It is employed for different types of Internet traffic including peer-to-peer (P2P) and content delivery traffic, and also for Internet traffic flow classification at a high speed

monitoring networks or managing QoE like it is described in Table 6. They treat estimation [81] or cross-validation [82] or Simple regression models. For video applications, QoE prediction can be performed by applying mathematical models based on QoS parameters [83], full-reference algorithms VQM [84], WFL based model [85]. Paper [86] shows the relationship between QoS (traffic characteristics) and QoE (linear, logarithmic or exponential). To evaluate the QoE, studies based on user engagement, (resiliency) abandonment rate, and frequency of visits are described in Ref. [87] and are using statistical tests (e.g., Pearson, Kendall). At the end, let's describe an interesting ML_QoE algorithm which employs supervised regression, in which the predictors are metrics (such as jitter, packet loss, rebuffering, startup delay, resolution) and the predicted outcome is the QoE score.

7.2 Adaptive coding

With the variation in the number of active users in SDN networks, we need sometimes to select the best codec and adapting it to delivered data. Voice over Internet Protocol (VoIP) is considered as much used protocol for voice service delivery. A fixed voice codec between a VoIP server and clients without consideration of the network conditions will cause inefficiency. To improve the VoIP quality, three factors should be considered: the optimized codec selection at first, second, VoIP flow control, and finally the consideration of network congestion. The authors of Ref. [88] propose the adaptive Mobile VoIP (mVoIP) service architecture based on network QoS parameters and predicted mVoIP QoE.

7.3 Flow classification and monitoring QoE

With increasing number of users and items, the network operators and service providers need methods to monitor the quality of the video services. The monitoring and prediction should be performed in real-time and in different parts of the network where subjective assessments are very hard to perform but objective measurement methods are done. New methods are required to help calculating QoE Key Performance Indicator (KPI) from QoS parameters. QoE monitoring fields know especially those approaches: flow classification, flow monitoring, location aware monitoring, and collaborative filtering. Flow Monitoring represents the method to determine the number and duration of re-buffering without access to user terminal.

7.4 Buffering impact

Network buffering takes place in hosts, switches and routers, impacts network performance and contribute to delays, jitter, and packet losses. The author of Ref. [89] presents a study on the impact of buffer sizes on QoE, in such applications: voice, video, and web browsing. The author provides also insights into the extension of QoE metrics for client-side error recovery. A hit rate analysis of caching schemes is performed to optimize Web QoE by focusing on YouTube video popularities.

7.5 Path optimization

In SDN, service providers have a complete control on their servers, networks and video players but no control done on access networks. Several researches deal with this problem. Zafar et al. [25] use a ML based traffic classification technique to identify path for each application types in SDN. Ali et al. [26] introduced MPLS-TE traffic engineering with OpenFlow. Saurav et al. [27] studied traffic aggregation in a packet circuit network. In the same context, authors of Ref. [90] propose to monitor streaming flow in real time and dynamically change routing paths using MPLS. They measure various video QoE metrics at video players running on user terminal in order to dynamically change routing paths.

This exhaustive state of the art about protocols and used techniques to enhance QoE in video streaming services lead us to propose an appropriate methods based on SSIM metric and Ml algorithms. Section 8 describes proposed method.

8. Simulation environment and testbed

Before presenting our approach based on SSIM and Machine learning algorithm over SDN networks, we have to put the environment, to describe topologies, scenarios and metrics to capture in order to best illustrate our approach. So, for this part of our paper, we choose to emulate some scenarios on SDN tools (Mininet and Open vSwitch) and experiment some politics in such environment. We will test some topologies making a variation in controller number, putting a Video streaming traffic with VLC server and measuring QoS parameters such as Jitter, RTT and bandwidth. The tests will be done on 2-min videos and ask users with web application to make a vote on those video qualities and generate a summary for this vote. Our politic to optimize end user quality of experience is to adapt a proposed flow to controller condition and topologies.

8.1 Testing on Mininet

For this part, we use Mininet and floodlight as an SDN simulation environment. We will make different topologies to evaluate the performance. So, in this section, we are going to measure three performance keys: Round-Trip Time, Jitter and Packet loss. Those measurements will be done on different topologies based on one Floodlight controller, two hosts and for four, six and eight switches for each case using UDP and TCP protocols. First of all, RTT (round-trip time) allows operators to understand the performance and help optimize their environment. Second, Jitter in IP networks represents the variation in the latency on a packet flow. Jitter is caused by congestion or path changes. It causes a hard problem in real-time communications. Finally, Packet loss is done when packets don't arrive to destination. This event can cause noticeable effects in all types of digital communications. In this section, we tried to make simple topologies composed of 4, 6 or 8 switches [(s1, ..., s4), (s1, ..., s6), (s1, ..., s8),] 2 hosts (h1, h2) and 1 controller (c0).

8.2 Graphical analyze

RTT: The round trip time has been increased when I increased the number of switches. Average $1 = 13,665$ ms < Average $2 = 14,229$ ms < Average $3 = 15,398$ ms.

Error rate: The average in the three experiences is 0.25% which is a very low average. It's true that when we increased the number of switches, the average has increased but it still has no influence in the transmission of UDP packets. The main reason of this loss is congestion. In fact, the UDP packet generator Iperf generates 10 Mbps which will be more than that by using the overhead of transmission (it is mainly equal to 11.9 Mbps after adding overheads). However, canal bandwidth is equal to 10 Mbps. These two facts generate the loss in the first switch. In conclusion, Mininet shows instability in its loss rate results but despite of that the loss rate is so low so Mininet provides as a good solution from the point of loss rate.

Jitter: Jitter also increased a little bit when the number of switches increased but it still low as a UDP packet transfer. Mininet didn't provide us with reasonable results. We have to use another platform in order to have stable variation in any topology, while implementing video stream service.

We put in the table below the performance of our simulated network for a 10MBtraffic, for next simulations, we attach a VLC server to the SDN controller and plan a 2-min video sequences and ask clients to vote for received quality in such networks. The testing is done on Mininet but if we had do it

Fig. 3 Performance evaluation.

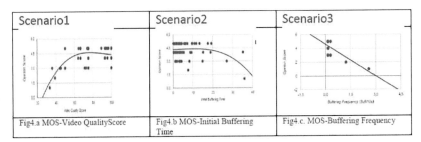

Fig. 4 Simulation results.

in real time network, we already took at different times of the day because this variation is crucial to control for the variability in the bandwidth which is affected by rush hours in real networks. We have chosen a 2-min duration video to ensure that the viewer does not get bored and leave the test. The chosen video also had clear visual properties and wide homogeneous textures, so that we can easily detect the variation of perception as a function of the quality's variation (Fig. 3).

8.3 Simulation scenarios

We run our three scenarios (Fig. 4) to make variation on Video Quality Score Initial Buffering Time and finally Buffering Frequency and measure each time the MOS obtained score.

Scenario 1: Mean Opinion Score as a function of Video Quality Score.

The VQS axis ranges from the first measured value which is about 29% to 95%. The value 29% is obtained from the lowest video quality the user might experience. The highest value on the scale is registered at 95% which can be translated in the fact that the user is watching the video at the maximum capacity of his terminal. The curve's shape shows a diminishing return on VQS. In other words, for values of VQS between 29% and 50%, the same improvement in VQS yields to a higher increase in opinion score compared to the region between 50% and 95%. In fact, users who are used to a poor VQS are likely to appreciate a small increase in its value which consequently leads to a higher rating. This is also translated on the portion when values are sup than 50% and above since the curve stabilizes and shows slight to no improvement in quality rating as a result of VQS increase.

Scenario 2: Mean Opinion Score as a function of Initial Buffering Time.

The initial buffering time takes values from 1 s up to 29% s. The curve shows a heavy concentration around the interval between the values of 1 and 5 s. This can be explained by the strategy followed by the video provider which prioritizes the minimization of the IBT. The curve can be described in two separate regions. In fact, over the first region (1–11 s) users' sensitivity to increasing IBT is not significant which can be translated in users' tolerance to a maximum waiting time of 11 s. However, on the second region (13–29 s) the opinion score curve decreases significantly as IBT increases; the user's experience becomes more and more unpleasant due to a higher waiting time.

Scenario 3: Mean Opinion Score as a function of Buffering Frequency.

The curve exhibits a downward sloping linear relationship between Opinion Score and Buffering Frequency per 8 s. The more stalls the viewer experiences while watching a video, the less immersion will be and the more annoying the experience gets.

8.4 Interpretation

The first and second curves taken together can infer to the fact that aiming for a minimum of 50% VQS and a maximum of 11 s in initial buffering time can maintain a high customer satisfaction. These numbers can be used as maximization and minimization constraints when optimizing the Quality of Experience. The third regression shows that buffering frequency is the one with the steepest linear relationship with the Opinion Score. This can be translated in the fact that no matter at what level of other KPIs the viewer is, any buffering is going to hurt his experience significantly. Therefore, buffering at the middle of the viewing experience is the factor hurting the experience the most for all users and should be given priority in optimizing the Quality of Experience.

9. Proposed approach: A ML-based enhancing QoE approach

9.1 SSIM approach

In this part, we investigate how Machine Learning approach works and how it can be used to improve quality of experience. We will start with an implemented work then we will expose our own solution. There are many Machine learning models of prediction. They aim to be able to predict the

QoE for end users without having to make subjective criteria for each new customer and media device. The work we will study involves the improvement of video quality experience through predicting SSIM measures: Structural Similarity. We expose the SSIM concept, some works from literature based on SSIM reference and Machine learning algorithms and finally, we let you discover our approach. SSIM is the measurement or prediction of image quality. It is based on an initial uncompressed or distortion-free image as reference. The idea of SSIM is to measure the structural similarity between the two images, rather than pixel to pixel difference. SSIM is an objective reference-based method to evaluate the quality of an image, which correlates well with the human perception and also allows an efficient computation. In the case of videos the SSIM is computed as the average SSIM over all the video frames. It is not easy to calculate the value of SSIM. So we will predict these measures through Machine Learning algorithms. Indeed, these algorithms are based primarily on a set of supervised learning techniques aimed at tackling discrimination and regression problems. First, unsupervised learning is used to extract an abstract representation of the raw data that capture the characteristics of the video. Then, supervised learning is performed to create a match between abstract representations and the corresponding SSIM coefficients of similar videos. Once the coefficients calculated, we determine the type of video encoding and the decision to take. So, in order to evaluate the objective QoE of a video, we will use the SSIM index, which is a complete reference measurement which measures the degradation of the image in terms of perception and of change in structural information.

SSIM is calculated by means of statistical parameters (mean, variance) calculated in a square window of size $N \times N$ (typically 8×8), which moves pixel by pixel in the entire image. The measures between the corresponding windows X and Y of the two images are calculated as follows:

$$SSIM(X, Y) = \frac{(2\mu_X\mu_Y + c_1)(2\sigma_{XY} + c_2)}{(\mu_X^2 + \mu_Y^2 + c_1)(\sigma_X^2 + \sigma_Y^2 + c_2)}$$

With μ: Mean and σ^2: variance.

The SSIM index goes from 0 to 1, where 0 represents the extreme case of completely different and 1 the case of completely identical pictures. We can make also the correspondence between SSIM and the Mean Opinion Score (MOS), which evaluates the subjective video quality perceived on a scale of 5 values from 1 (poor) to 5 (excellent). So SSIM index 1 corresponds to excellent quality and therefore an index 5 scale MOS as shows Table 11.

Table 11 MOS.

SSIM	MOS	Quality	Impairment
≥ 0.99	5	Excellent	Imperceptible
[0.95, 0.99]	4	Good	Perceptible but not annoying
[0.88, 0.95]	3	Fair	Slightly annoying
[0.5, 0.88]	2	Poor	Annoying
<0.5	1	Bad	Very annoying

Since the calculation of the SSIM can be difficult or impossible, we will use Machine Learning algorithms to predict the quality of the video that the user perceives subjectively. We will use then a data set of videos with their approximate SSIM index to teach the machine. So that it will be able to predict and calculate the quality of the video. Once the quality is calculated, we are going to use a set of optimization and operational research algorithms needed to choose the right encoding for the video dependent on its quality: this mechanism is called VAC: Video Admission Control. These algorithms have already been tested before. We have the possibility to double-check their performances. But, we will focus on studying the performance of our improved algorithm.

9.2 SSIM based works

With the rapid increase of streaming multimedia applications, there has been a strong and complex demand of Quality-of-Experience (QoE) measurement and QoE-monitoring technologies. Therefore, many techniques have been developed in order to evaluate as correctly as possible this perceptual quality. To inquire QoE measurement and predictions, most existing methods are based on the presentation quality of compressed video and objectives metrics such as PSNR, SSIM, MS-SSIM, etc., while others are interested on Machine Learning to predict the suitable user perception in real time based on previous measurements.

Rehman et al. [91,92] propose a full-reference video QoE measure, SSIMplus that provides real-time prediction of the video quality based on human visual system behaviors, display device properties (such as screen size, resolution, and brightness) video content characteristics, and viewing conditions (such as viewing distance and angle). We compared the performance of the proposed algorithm to the most popular and widely used full reference VQA measures that include Peak Signal-to-Noise Ratio (PSNR), Structural

Similarity (SSIM), Multiscale Structural Similarity (MS-SSIM), etc. They conclude that their algorithm is much better and faster tool than other traditional methods. Besides Lievens et al. [93] reported that the classical quality measurements (PSNR, SSIM, VQM) failed to predict the perceived video quality in HTTP adaptive streaming due to quality fluctuations and proposed an empirical quality metric to account for the streaming-specific distortions. Moreover, Thomas et al. [94] are based on the mapping method between SSIM, VQM and QoE to conduct an extensive measurement framework to study the influence of resolution of the video, scaling method, frame rate and video content types on the QoE. In other work [95] User experience is objectively expressed in terms of the average structural similarity (SSIM) index to analyze the video quality, authors propose a 4-degree polynomial approximation of the SSIM as a function of the coded video rate, they found that, measuring the QoE of a video in terms of SSIM, and expressing it as a function of the logarithm of the source rate, normalized to the full quality rate, it is possible to accurately approximate the quality vs rate distortion curve of the video as a 4-degree polynomial function. The polynomial coefficients can be tagged to the video and used for QoE-aware resource allocation. From their experiments they conclude that videos exhibit a relatively large SSIM (>0.94) even when the transmission rate is decreased to 1% of their full quality rate while others suffer a quick fall of SSIM already when the rate is reduced to 10% of the full rate. They conclude also that videos with similar SSIM attitude appear widely homogeneous in terms of scenes dynamics, with most dynamic videos exhibiting a quicker drop of SSIM as the RSF decreases. In Refs. [96,97] authors use some objective parameters such as SSIM, PSNR, NIQE, etc. to evaluate the influence of bitrates and stalling on the quality of experience, those metrics allow them to give an idea about the user's opinion.

9.3 Machine learning based works

Other researchers use another method for estimating QoE for multimedia services based on Machine Learning algorithms, these algorithms give good results in real time; for example, Charonyktakis et al. [98] propose the MLQoE, a modular algorithm for user-centric QoE prediction. This framework employs multiple machine learning (ML) algorithms, namely, Artificial Neural Networks, Support Vector Regression machines, Decision Trees, and Gaussian Naive Bayes classifiers, and tunes their hyper-parameters for VoIP service. It uses also the Nested Cross Validation method to select the best classifier and the corresponding best hyper-parameter values and predicts the performance of the final model. This model predicts accurately the QoE score. Specifically, a mean absolute error of <0.5 and median absolute error of <0.30 can be achieved. In other hand, Sajid Mushtaq et al. [99] have been achieved the correlation between QoS and QoE in search of capturing the degree of user opinion, they are based on Machine Learning to determine the most suitable one for the task of QoS/QoE correlation in order to study the effect of QoS metric on the QoE to deliver a better quality of service to end-users. This work evaluated six classifiers and determines the most suitable one for the task of QoS/QoE correlation. Experimental results show the in case of mean absolute error rate, it is observed that DT has a good performance as compared to all other algorithms. Also in Ref. [100], a machine learning technique is proposed using a subjective quality feedback. This technique is used to model dependencies of different QoS metrics related to network and application layer to the QoE of the network services and summarized as an accurate QoE prediction model. Finally, authors in Ref. [101] used Machine Learning for estimating audiovisual quality multimedia service. They used to train the models Random Forests and Multi-Layer Perceptron methods, results show that Random Forests based methods outperform Multi-Layer Perceptron methods in terms of RMSE, Pearson Correlation coefficient value and 95% confidence interval boundaries.

9.4 Proposed approach

We saw in the previous section, an example of an algorithm of Machine Learning to ameliorate QoE. Indeed, this algorithm when deciding at the end to choose the type of video encoding it only takes the quality expected by the user into account. But we can consider this approach as limited because we still have bandwidth problems, network congestion and other services that are running and that may affect the network quality and thus reduce the quality of experience and cause a cut during video streaming.

Now, we are going to detail our proposal of improvement to this algorithm to further enhance the user's quality of experience. As we have already explained in the SSIM part: this solution is very interesting. Indeed, depending on the SSIM index, we will know the minimum required quality level and therefore the appropriate codec to encode the video and send it over the network.

But this choice of codec ignores the state of the network, limitations of bandwidth and anything can happen on the network. It only takes the minimum quality of the video into account. But, sometimes this minimum is not even valid for the user's network conditions and could provoke the cut of the video streaming which implies a very low QoE.

Our improvement addresses these network conditions. So, after calculating the quality of the video, we will not directly choose the codec to use. Using the SDN centralized architecture, we can predict the state of the network and know the different metrics that can influence on the transmission of the video and that can cause poor quality of experience. After calculating these measures, we will calculate the correlation between the SSIM index and network quality index. This will generate a new index from which we will choose the appropriate codec and so we will ensure a minimum QoE while taking into account two conditions: the state of the network and the transmission part as well as the minimum quality required for the video.

All this work is going to be in SDN architecture since we are going to control the network with the need of the controller. Then, as we have already explained, it is almost impossible to calculate the SSIM indicator, we can only predict thanks to Machine Learning algorithms, so we implemented these prediction algorithms and chose a free data set not very long videos (about 1 min and a half) to teach our machine to calculate the SSIM. Once calculated: this index gives us an idea about the minimum quality required for a video. All this work is done in the video streaming server which is responsible for the delivery of videos. The SSIM algorithm is already implemented and is open source so we downloaded it and we changed it. The algorithm in its original version calculates the similarity between two consecutive images. We will modify it to calculate an index for the entire video. Once the SSIM is calculated for the video, we need to calculate the state of the network to select the appropriate codec. It can be collected from the network using QoS metrics or predicted depending on the previous state. After that, we will calculate a new index that is the correlation between SSIM index and the state of the network. This index is the way to select the appropriate codec. Once the codec is selected and the video is being streaming and since that the network parameters may change and streaming video is a real-time service, the choice of codec will also be dynamic and can change from one moment to another.

We can see from Table 12 that each index is associated with a particular codec according to the algorithm already presented in the previous section. This codec will vary over time and will depend on network state in real-time.

Table 12 Chosen codec.

Index	Chosen codec
$0.66 \rightarrow 0.99$	x.264
$0.33 \rightarrow 0.66$	H.264
$0.00 \rightarrow 0.33$	MPEG4

9.5 Experiments

In this section, we will deploy in our test platform several scenarios and evaluate our solution's performance. Our platform consists of two levels: First, the data plane that is the switches and the transmission part and then, the control plane, which consists of controllers and the commanding part. The QoE measure we will use for evaluating network performance is the Success rate: the percentage of packets received by the host compared to total packets exchanged. This measure will be then correlated with the SSIM index for selecting the best codec. To study the performance of this proposal, it is necessary to use different simulation scenarios. Indeed, we will test our platform for a real-time service that is video streaming.

9.5.1 First scenario

After deploying the entire architecture, now we'll just start with a single user who will connect to our streaming video service, it will try to watch all the videos that exist in our Streaming video server without deploying our solution. It will therefore seek switches to reach the server. The number of nodes that we used in this first simulation is 10 and the user is only going to watch from the database of videos. We have 15 videos and we will restrict our study to the percentage of packets received successfully. In Table 13, we summarized the different results of the user's access to our server for the 15 videos. We can notice that we have a situation here where the user is logged alone and therefore enjoys all of the bandwidth. So we have an average of 97% of success. This is due mainly to the fact that the user is alone and there is no congestion or other services launched at the same time to clutter nodes and decrease the quality perceived by the user.

9.5.2 Second scenario

In this second scenario, we will increase the number of nodes and the number of users requesting the same service at the same time and all trying to have streaming video service without implementing our idea and therefore this scenario will be seen as an extreme case scenario or saturation case. So

Table 13 Simulation results.

	First scenario			Second scenario			Third scenario	
Video	Duration time	Percentage of success packets	Video	Duration time	Percentage of success packets	Video	Duration time	Percentage of success packets
Video 1	1:35	95.21%	Video 1	1:35	44.25%	Video 1	1:35	78.81%
Video 2	1:28	96.57%	Video 2	1:28	46.19%	Video 2	1:28	79.84%
Video 3	1:06	98.48%	Video 3	1:06	48.14%	Video 3	1:06	85.15%
Video 4	1:52	95.32%	Video 4	1:52	45.68%	Video 4	1:52	75.72%
Video 5	1:26	98.21%	Video 5	1:26	46.43%	Video 5	1:26	81.26%
Video 6	1:32	97.45%	Video 6	1:32	47.76%	Video 6	1:32	79.16%
Video 7	1:41	97.24%	Video 7	1:41	45.17%	Video 7	1:41	77.54%
Video 8	1:22	98.17%	Video 8	1:22	47.64%	Video 8	1:22	82.69%
Video 9	1:17	97.40%	Video 9	1:17	47.45%	Video 9	1:17	82.81%
Video 10	1:09	99.04%	Video 10	1:09	48.32%	Video 10	1:09	84.85%
Video 11	1:27	97.51%	Video 11	1:27	45.81%	Video 11	1:27	81.44%
Video 12	1:19	98.27%	Video 12	1:19	47.55%	Video 12	1:19	83.07%
Video 13	1:41	96.24%	Video 13	1:41	45.54%	Video 13	1:41	78.74%
Video 14	1:22	97.04%	Video 14	1:22	48.71%	Video 14	1:22	82.18%
Video 15	1:52	95.89%	Video 15	1:52	46.86%	Video 15	1:52	74.98%

we will use 40 switches and 10 users who all want to connect to the server at the same time and we will look to a single user and follow his trial to connect to the 15 videos.

In Table 13, we summarized the different results of the user's access to our architecture for the 15 videos in this second scenario. Through the results, we can deduce that in a scenario of saturation or congestion when everyone wants to connect to the same server through the same resources, the quality perceived by the user decreases and this is confirmed by the average of 47% for the success rate of packets.

9.5.3 Third scenario

In the third scenario the same number of users trying to have that same service with a user who will try to watch the 15 streaming videos and nodes that provide this service implements the ameliorated SSIM solution and this will give us an idea on the behavior of our approach when there is a streaming service requested simultaneously for a normal user and a user implementing our solution.

In Table 13 we summarized the different results of the user's access to our architecture for 15 videos after deploying our algorithm. We can easily notice how the percentage rose again to stabilize in the 83% rate of success of exchanged packets: this rate is better than the previous rate and in the same network conditions could lead us to conclude that our solution is efficient and improves the user's QoE.

The graphic below shows the difference between the three scenarios (Fig. 5).

9.5.4 Results

We believe in the future we can test more factors and ameliorate our algorithm to consider single metric to be able to improve the quality of the experience. We also think of adding and improving our proposed solution so that it will have a better performance for other services such as video conference and real-time applications.

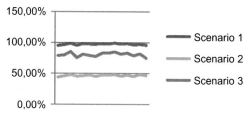

Fig. 5 Users' access.

We will also look at other approaches. As we rather concentrated in this work on machine learning approach that leaves us even more prospects and opportunities to perform in the future to finally combine all solutions.

10. Conclusion, perspectives and futures work

Preliminary experimental results show that Mininet can effectively estimate the efficiency of adaptive solution and intelligent placed on controllers. This work allows claiming that our proposed approach is promising, as the simulation procedure for the QoE calculation can be efficiently performed. Further experimental research is needed to estimate the efficiency of this approach for different types of scenarios and their parameters. We argue that Mobile web platform is needed to allow the Internet service providers to collect these measures, analyze them and draw recommendations to improve the degree of their user's satisfaction. In this work, we proposed and worked on a novel machine learning algorithm for objectively estimating QoE service based on SSIM parameters. Further simulations will be carried out on the suitability of both Mininet and Open vSwitch to calculate and predict QoE for video streaming services. In particular, this paper focuses on a QoE overview and discusses QoE-metrics and video streaming protocols. We propose an algorithm based on SSIM and Machine learning technique to enhace QoE in SDN networks. In the future, the performance of the proposed algorithm will be evaluated in real networks with more than one simulation environment.

References

[1] A. Bentaleb, A.C. Begen, R. Zimmermann, S. Harous, SDNHAS: an SDN-enabled architecture to optimize QoE in HTTP adaptive streaming, IEEE Trans. Multimedia 19 (10) (2017) 2136–2151.
[2] C.-C. Wu, K.-T. Chen, I.E.E.E. Yu-Chun Chang, C.-L. Lei, Crowdsourcing multimedia QoE evaluation: a trusted framework, IEEE Trans. Multimedia 15 (5) (2013) 1121.
[3] J. Song, F. Yang, Y. Zhou, S. Wan, H.R. Wu, QoE evaluation of multimedia services based on audiovisual quality and user interest, IEEE Trans. Multimedia 18 (3) (2016) 444–457.
[4] N. Staelens, S. Moens, W. Van den Broeck, I. Mariën, B. Vermeulen, Assessing quality of experience of IPTV and video on demand services in real-life environments, IEEE Trans. Broadcast. 56 (4) (2010) 458–466.
[5] K.U.R. Laghari, N. Crespi, K. Connelly, Toward total quality of experience: a QoE model in a communication ecosystem, IEEE Commun. Mag. 50 (4) (2012) 58–65.
[6] W.C. Hardy, QoS Measurement and Evaluation of Telecommunications Quality of Service, John Wiley & Sons, Chichester, UK, 2001.
[7] J. HENS, J. Caballero, Triple Play: Building the Converged Network for IP, VoIP and IPTV, Wiley, England, 2008. ISBN 978-047-0753-675.

[8] ITU –T Recommendation G.1080, Quality of Experience Requirements for IPTV Services, 2008.
[9] F. Agboma, A. Liotta, Quality of experience management in mobile content delivery systems. Telecommun. Syst. 49 (2012) 85, https://doi.org/10.1007/s11235-010-9355-6.
[10] R. Stankiewicz, P. Cholda, A. Jajszczyk, QoX: what is it really? IEEE Commun. Mag. 49 (4) (2011) 148–158, https://doi.org/10.1109/MCOM.2011.5741159.
[11] F. Kuipers, et al., Techniques for measuring quality of experience, in: Proceedings of the 8th International Conference on Wired/Wireless Internet Communications, 2010, pp. 216–227. ISBN 978-3-642-13314-5.
[12] R.E. Mayer, Multimedia Learning, Cambridge University Press, New York, 2001.
[13] K. De Moor, I. Ketyko, W. Joseph, T. Deryckere, L. De Marez, L. Martens, G. Verleye, Proposed framework for evaluating quality of experience in a mobile, testbed-oriented living lab setting. Mob. Netw. Appl. 15 (2010) 378, https://doi.org/10.1007/s11036-010-0223-0.
[14] N.R. Veeraragavan, H. Meling, R. Vitenberg, QoE estimation models for tele-immersive applications, in: EUROCON, 2013 IEEE, 2013. 1–4 July.
[15] S. Moller, C. Wai-Yip, N. Cote, T.H. Falk, A. Raake, M. Waltermann, Speech quality estimation: models and trends, IEEE Signal Process. Mag. 28 (6) (2011) 18–28. Institute of Electrical and Electronic Engineers.
[16] I. Avcibas, B. Sankur, K. Sayood, Statistical evaluation of image quality measures, J. Electron. Imaging 11 (2002) 206–223.
[17] S. Chikkerur, V. Sundaram, M. Reisslein, L. Karam, Objective video quality assessment methods: a classification, review, and performance comparison, IEEE Trans. Broadcast. 57 (2) (2011) 165–182.
[18] G. Cofano, L. De Cicco, T. Zinner, A. Nguyen-Ngoc, P. Tran-Gia, S. Mascolo, Design and experimental evaluation of network-assisted strategies for HTTP adaptive streaming, in: 7th ACM Conference MMSys, 2016.
[19] M.S. Mushtaq, B. Augustin, A. Mellouk, Empirical study based on machine learning approach to assess the QoS/ QoE correlation, in: 17th European Conference on Networks and Optical Communications (NOC), 2012, pp. 1–7.
[20] P. Georgopoulos, Y. Elkhatib, M. Broadbent, M. Mu, N. Race, Towards network-wide QoE fairness using openflow-assisted adaptive video streaming, in: Proc. ACM SIGCOMM Workshop on Future Human-Centric Multimedia Networking, 2013.
[21] S. Petrangeli, J. Famaey, M. Claeys, S. Latré, F. De Turck, QoE-driven rate adaptation heuristic for fair adaptive video streaming. ACM Trans. Multimed. Comput. Commun. Appl. 12 (12) (2016) 28, https://doi.org/10.1145/2818361.
[22] A. Bentaleb, A.C. Begen, R. Zimmermann, "SDNDASH: Improving QoE of HTTP Adaptive Streaming Using Software Defined Networking", MM '16, Proceedings of the 24th ACM International Conference on Multimedia, Amsterdam, Netherlands, October 15–19 (2016) 1296–1305.
[23] S. Ramakrishnan, X. Zhu, F. Chan, K. Kambhatla, SDN based QoE optimization for HTTP-based adaptive video streaming, in: IEEE International Symposium on Multimedia Miami, FL, USA, 2015.
[24] A. Takahashi, H. Yoshino, N. Kitawaki, Perceptual QoS assessment technologies for VoIP, IEEE Commun. Mag. 42 (7) (2004) 28–34.
[25] Z.A. Qazi, J. Lee, T. Jin, G. Bellala, M. Arndt, G. Noubir, Application-awareness in SDN, in: Proceedings of the ACM SIGCOMM Conference, Hong Kong, China, 2013.
[26] A.R. Sharafat, S. Das, G. Parulkar, N. McKeown, MPLS-TE and MPLS VPNS with openflow, in: Proceedings of the ACM SIGCOMM Conference, Toronto, Ontario, Canada, 2011.
[27] S. Das, Y. Yiakoumis, G. Parulkar, N. McKeown, P. Singh, D. Getachew, P. Desai, Application-aware aggregation and traffic engineering in a converged packet-circuit

network, in: Optical Fiber Communication Conference and Exposition (OFC/NFOEC) and the National Fiber Optic Engineers Conference, Los Angeles, USA, 2011.

[28] About NOX, [Online]. Available: http://www.noxrepo.org/.

[29] M. Jarschel, F. Wamser, T. Hohn, T. Zinner, P. Tran-Gia, SDNbased application-aware networking on the example of youtube video streaming, in: 2nd European Workshop on Software Defined Networks (EWSDN), Berlin, Germany, 2013.

[30] K.J. Ma, R. Bartos, S. Bhatia, R. Naif, Mobile video delivery with HTTP. IEEE Commun. Mag. 49 (4) (2011) 166–175, https://doi.org/10.1109/MCOM.2011.5741161.

[31] O. Oyman, S. Singh, Quality of experience for HTTP adaptive streaming services. IEEE Commun. Mag. 50 (4) (2012) 20–27, https://doi.org/10.1109/MCOM.2012.6178830.

[32] S. Ouellette, L. Marchand, S. Pierre, A potential evolution of the policy and charging control/QoS architecture for the 3GPP IETF-based evolved packet core. IEEE Commun. Mag. 49 (5) (2011) 231–239, https://doi.org/10.1109/MCOM.2011.5762822.

[33] M. Eckert, T.M. Knoll, T. Bauschert (Ed.), QoE Management Framework for Internet Services in SDN Enabled Mobile Networks, IFIP International Federation for Information Processing, 2013, pp. 112–123. EUNICE 2013, LNCS 8115. 2013.

[34] K. Kim, Y.J. Choi, Performance comparison of various VoIP codec in wireless environments, in: Proc. ACM International Conference on Ubiquitous Information Management and Communication (ICUIMC 11), Korea, 2011.

[35] R. Schatz, T. Hossfeld, P. Casas, Passive youtube QoE monitoring for isps, in: Proc. of IMIS, 2012.

[36] M. Eckert, T.M. Knoll, F. Schlegel, Advanced MOS calculation for network based QoE estimation of TCP streamed video services, in: Proc. of ICSPCS, 2013.

[37] T. De Pessemier, K. De Moor, W. Joseph, L. De Marez, L. Martens, Quantifying the influence of rebuffering interruptions on the user's quality of experience during mobile video watching, IEEE Trans. Broadcast. 59 (1) (2013) 47–61.

[38] Z. Su, Q. Xu, Q. Qi, Big Data in Mobile Social Networks: A QoE-Oriented Framework, IEEE Network, 2016. Jan/Feb.

[39] V.K. Adhikari, M. Varvello, V. Hilt, M. Steiner, Z.-L. Zhang, Unreeling netflix: understanding and improving multi-CDN movie delivery, in: Proceedings of the 2012 IEEE INFOCOM, 2012, pp. 1620–1628.

[40] T. Koponen, M. Chawla, B.-G. Chun, A. Ermolinskiy, K.H. Kim, S. Shenker, I. Stoica, A data-oriented (and beyond) network architecture, in: Proc. ACM SIGCOMM'07, 2007.

[41] P. Brooks, B. Hestnes, User measures of quality of experience: why being objective and quantitative is important, IEEE Netw. 24 (2) (2010) 8–13.

[42] C. Ge, N. Wang, G. Foster, M. Wilson, Toward QoE-assured 4K video-on-demand delivery through mobile edge virtualization with adaptive prefetching, IEEE Trans. Multimedia 19 (10) (2017) 2222–2237.

[43] H. Al-Zubaidy, V. Fodor, G. Dán, M. Flierl, Reliable video streaming with strict playout deadline in multihop wireless networks, IEEE Trans. Multimedia 19 (10) (2017) 2238–2251.

[44] X. Zhu, S. Mao, M. Hassan, H. Hellwagner, Guest editorial: video over future networks, IEEE Trans. Multimedia 19 (10) (2017) 2133–2135.

[45] A. Rao, A. Legout, Y.-s. Lim, D. Towsley, C. Barakat, W. Dabbous, Network characteristics of video streaming traffic, in: Proceedings of the CoNEXT, 2011.

[46] N. Kushik, J. Pokhrel, N. Yevtushenko, A.R. Cavalli, W. Mallouli, QoE prediction for multimedia services: comparing fuzzy and logic network approaches, IJOCI 4 (3) (2014) 44–64. Felipe Lalanne.

[47] J. Hua, Study on the application of rough sets theory in machine learning, in: Intelligent Information Technology Application, 2008. IITA '08. Second International Symposium on, 1 2008, pp. 192–196.

[48] K. Laghari, I. Khan, N. Crespi, Quantitative and qualitative assessment of QoE for multimedia services in wireless environment, in: Proceedings of the 4th Workshop on Mobile Video, 2012.

[49] J. Pokhrel, B. Wehbi, A.N.P. Morais, A.R. Cavalli, E. Allilaire, Estimation of QoE of video traffic using a fuzzy expert system, in: CCNC, 2013, pp. 224–229.

[50] N. Kushik, N. Yevtushenko, A.R. Cavalli, W. Mallouli, J. Pokhrel, QoE estimation for web service selection using a fuzzy-rough hybrid expert system, in: AINA, 2014, pp. 629–634.

[51] P. Calyam, P. Chandrasekaran, G. Trueb, N. Howes, R. Ramnath, D. Yu, L. Ying, L. Xiong, D. Yang, Multi-resolution multimedia QoE models for IPTV applications, Int. J. Digit. Multimed. Broadcast. 2012 (2012) 904072.

[52] S. Mohamed, G. Rubino, A study of real-time packet video quality using random neural networks, IEEE Trans. Circuits Syst. Video Technol. 12 (12) (2002) 1071–1083.

[53] A. Khan, L. Sun, E. Ifeachor, J. Fajardo, F. Liberal, H. Koumaras, Video quality prediction models based on video content dynamics for H. 264 video over UMTS networks. Int. J. Digit. Multimed. Broadcast. 2010 (2010) 608138, https://doi.org/10.1155/2010/608138.

[54] N. Kushik, N. Yevtushenko, A.R. Cavalli, W. Mallouli, J. Pokhrel, Evaluating web service QoE by learning logic networks, in: WEBIST, 2013, pp. 168–176.

[55] G. Rubino, P. Tirilly, M. Varela, Evaluating users satisfaction in packet networks using random neural networks, in: Artificial Neural Networks, ICANN 2006, Volume 4131 of Lecture Notes in Computer Science, Springer, Berlin, 2006, pp. 303–312.

[56] D. Kwon, R. Thay, H. Kim, H. Ju, QoE-based adaptive mVoIP service architecture in SDN networks, ICN 2014, in: The Thirteenth International Conference on Networks, 2014.

[57] Yao Liu, Sujit Dey, Don Gillies, Faith Ulupinar, Michael Luby, User Experience Modeling for DASH Video, 20th International Packet Video Workshop, Dec 12–13, 2013, https://doi.org/10.1109/PV.2013.6691459.

[58] A. Sfairopoulou, Bellalta, Macian, How to tune VoIP codec selection in WLANs? IEEE Commun. Lett. 12 (8) (2008) 551–553.

[59] B. Bellalta, C. Macian, A. Sfairopoulou, C. Cano, Evaluation of joint admission control and VoIP codec selection policies in generic multirate wireless networks, in: Proc. IEEE International Conference on Next Generation Teletraffic and Wired/Wireless Advanced Networking (NEW2AN 07) St. Peterburg, Russia, 2007, pp. 342–355.

[60] S.L. Ng, S. Hoh, D. Singh, Effectiveness of adaptive codec switching VoIP application over heterogeneous networks, in: Proc. International Conference on Mobile Technology, Application and Systems, 2005. China, 15–17 Nov.

[61] O. Hohlfeld, Impact of Buffering on Quality of Experience, Oliver Hohlfeld, 2013.

[62] R. Pantos, HTTP Live Streaming, 2011, IETF, http://tools.ietf.org/html/draft-pantos-http-live-streaming-06.

[63] Sandvine, Global Internet Phenomena Report, Sandvine Incorporated ULC, 2011.

[64] C. Liu, I. Bouazizi, M. Gabbouj, Rate adaptation for adaptive HTTP streaming, in: Proceedings of the Second Annual ACM Conference on Multimedia Systems (MMSys), 2011.

[65] The Information Sciences Institute (ISI), Network Simulator Ns-2, [Online], last update 2011. Available http://isi.edu/nsnam/ns/.

[66] D. Jarnikov, T. Ozcelebi, Client intelligence for adaptive streaming solutions, in: Proceedings of the IEEE International Conference on Multimedia and Expo (ICME), 2010.

[67] S. Akhshabi, A.C. Begen, C. Dovrolis, An experimental evaluation of rate-adaptation algorithms in adaptive streaming over HTTP, in: Proceedings of the Second Annual ACM Conference on Multimedia Systems (MMSys), 2011.

[68] M. Zink, O. Kunzel, J. Schmitt, R. Steinmetz, Subjective impression of variations in layer encoded videos, in: Proceedings of the 11th International Conference on Quality of Service (IWQoS), 2003.

[69] K. Miller, E. Quacchio, G. Gennari, A. Wolisz, Adaptation algorithm for adaptive streaming over HTTP. in: 19th International Packet Video Workshop (PV), May 10–11, 2012, IEEE, https://doi.org/10.1109/PV.2012.6229732.

[70] iDASH—Yago Sánchez MMSYS-2011, ppt, fraunhofer.

[71] A. El Essaili, D. Schroeder, D. Staehle, M. Shehada, W. Kellerer, E. Steinbach, Quality-of-experience driven adaptive HTTP media delivery, in: IEEE International Conference on Communications ICC, 2013.

[72] S. Hemami, A. Reibman, No-reference image and video quality estimation: applications and human-motivated design, Signal Process. Image Commun. 25 (7) (2010) 469–481.

[73] M. Ries, O. Nemethova, M. Rupp, Video quality estimation for mobile H.264/AVC video streaming, J. Commun. 3 (1) (2008) 41–50.

[74] R. Mok, E. Chan, Measuring the quality of experience of HTTP video streaming, in: Proceeding of 2011 IEEE International Symposium on Integraed Network Magnagement, Dublin, 2011.

[75] J. Jiang, V. Sekar, H. Zhang, Improving fairness, efficiency, and stability in HTTP-based adaptive video streaming with festive, in: Proceedings of the ACM 8th International Conference on Emerging Networking Experiments and Technologies, 2012.

[76] T. Abar, A. Ben Letaifa, S. Elasmi, Enhancing QoE Based On Machine Learning and DASH in SDN Networks, Accepted in WAINA18, Cracow, Poland, May 2018, IEEE.

[77] E. Thomas, M. van Deventer, T. Stockhammer, A.C. Begen, J. Famaey, Enhancing MPEG DASH performance via server and network assistance. SMPTE Motion Imaging J. 126 (1) (2017) 22–27, Stevenhage: IET, https://doi.org/10.5594/JMI.2016.2632338.

[78] Y. Liu, S. Dey, D. Gillies, F. Ulupinar, M. Luby, User Experience Modeling for DASH Video. IEEE, 2013, https://doi.org/10.1109/PV.2013.6691459.

[79] P. Ni, R. Eg, A. Eichhorn, Flicker effects in adaptive video streaming to handheld devices, in: Proceedings of the ACM.International Multimedia Conference (ACM MM), 2011.

[80] T.T. Nguyen, G. Armitage, A survey of techniques for internet traffic classification using machine learning, IEEE Commun. Surv. Tutorials 10 (4) (2008) 56–76.

[81] V. Menkovski, A. Oredope, A. Liotta, A.C. S'anchez, Optimized online learning for QoE prediction, in: Proceedings 21st Benelux Conference on Artificial Intelligence, 2009.

[82] V. Menkovski, G. Exarchakos, A. Liotta, Online QoE prediction, in: 2010 Second International Workshop on Quality of Multimedia Experience, IEEE, 2010.

[83] P. Reichl, S. Egger, R. Schatz, A. D'Alconzo, The logarithmic nature of QoE and the role of the weber-fechner law in QoE assessment, in: 2010 IEEE International Conference on Communications, 2010.

[84] M.H. Pinson, S. Wolf, A new standardized method for objectively measuring video quality broadcasting. IEEE Trans. Broadcast. 50 (3) (2004) 312–322, https://doi.org/10.1109/TBC.2004.834028.

[85] T. Hoßfeld, S. Egger, R. Schatz, M. Fiedler, K. Masuch, C. Lorentzen, Initial delay vs. interruptions: between the devil and the deep blue sea, in: 2012 Fourth International Workshop on Quality of Multimedia Experience, 2012.

[86] J. Shaikh, M. Fiedler, D. Collange, Quality of experience from user and network perspectives. Ann. Telecommun. 65 (2010) 47, https://doi.org/10.1007/s12243-009-0142-x.

[87] S.S. Krishnan, R.K. Sitaraman, Video stream quality impacts viewer behavior: inferring causality using quasi-experimental designs. IEEE/ACM Trans. Networking 21 (6) (2013) 2001–2014, https://doi.org/10.1109/TNET.2013.2281542.

[88] D. Kwon, R. Thay, H. Kim, J. Hongtaek, QoE-based adaptive mVoIP service architecture in SDN networks, in: ICN 2014: The Thirteenth International Conference on Networks, 2014.

[89] Oliver Hohlfeld, Impact of Buffering on Quality of Experience, Thesis report (n.d.).

[90] H. Nam, K.-H. Kim, J.Y. Kim, H. Schulzrinne, Towards QoE-Aware Video Streaming Using SDN, Globcom, 2014.

[91] A. Rehman, K. Zeng, Z. Wang, Display device-adapted video quality-of-experience assessment, in: IS&T-SPIE Electronic Imaging, Human Vision and Electronic Imaging XX, 2015.

[92] The SSIMplus Index for Video Quality-of-Expeirnce Assessment, http://ece.uwaterloo.ca/z70wang/research/ssimplus.

[93] J. Lievens, A. Munteanu, D. De Vleeschauwer, W. Van Leekwijck, Perceptual video quality assessment in HTTP adaptive streaming, in: IEEE Int. Conf. on Consumer Electronics (ICCE), 2015, pp. 72–73. Las Vegas, NV, Jan. 9–12.

[94] T. Zinner, O. Abboud, O. Hohlfeld, T. Hossfeld, Impact of frame rate and resolution on objective QoE metrics, in: Second International Workshop on Quality of Multimedia Experience, 2010.

[95] M. Zanforli, D. Munarett, A. Zanell, M. Zorzi, "SSIM-based video admission control and resource allocation algorithms", modeling and optimization in mobile, Ad Hoc, and wireless networks (WiOpt), in: 2014 12th International Symposium, Hammamet, Tunisia, 12–16, 2014.

[96] K. Zeng, H. Yeganeh, Z. Wang, In: Quality-of-Experience of Streaming Video: Interactions Between Presentation Quality and Playback Stalling, IEEE International Conference on Image Processing (ICIP), Phoenix, AZ, USA, 2016.

[97] Z. Duanmu, K. Zeng, K.M.A. Rehman, Z. Wang, A quality-of-experience index for streaming video. IEEE J. Sel. Top. Sign. Proces. 11 (1) (2016) 154–166, https://doi.org/10.1109/JSTSP.2016.2608329.

[98] P. Charonyktakis, M. Plakia, I. Tsamardinos, M. Papadopouli, On user-centric modular QoE prediction for VoIP based on machine-learning algorithms, IEEE Trans. Mob. Comput. 15 (2016) 1443–1456.

[99] M. Sajid Mushtaq, B. Augustin, A. Mellouk, Empirical study based on machine learning approach to assess the QoS/QoE correlation, in: 17th European Conference on Networks and Optical Communications (NOC), Vilanova i la Geltru, Spain, 2012.

[100] V. Menkovski, G. Exarchakos, A. Liotta, Machine learning approach for quality of experience aware networks, in: Intelligent Networking and Collaborative Systems (INCOS), 2010 2nd International Conference on Nov, 2010.

[101] E. Demirbilek, J.-C. Gr'egoire, Towards Reduced Reference Parametric Models for Estimating Audiovisual Quality in Multimedia Services, in: 2016 IEEE International Conference on Communications, 2016.

About the author

 **Dr. Asma Ben Letaifa** is an assistant professor and member of the MEDIATRON research lab at the Higher School of Communications, SUPCOM, University of Carthage, Tunisia. She holds a Telecom Engineering Degree from SUPCOM, University of Carthage, Tunisia and a PhD jointly from SUPCOM, University of Carthage and UBO, Université de Bretagne Occidentale, Brest, France.

Her research activities focus on telecom services, cloud and mobile cloud architectures, service orchestration, bigData and quality of experience in an SDN/NFV environment. She is author and coauthor of several articles on these subjects.

She is also author of several courses on telecommunications services, network modeling with queuing theory, web content, cloud architectures and virtualization, massive BigData content and machine learning algorithms. She is also coauthor of the "Linux practices" MOOC on the FUN platform.